scenic science of the national parks
O9-BTJ-571

# scenic science of the national parks

## an explorer's guide to wildlife, geology, and botany

EMILY HOFF & MAYGEN KELLER

ILLUSTRATIONS BY JILLIAN BARTHOLD

TEN SPEED PRESS
California | New York

# contents

introduction 1
- how to use this book 3
- why we wrote this book 4
- land recognition 5
- safety in the parks 6

## PACIFIC NORTHWEST

**olympic • OLYM** 9
**mount rainier • MORA** 17
**crater lake • CRLA** 25
**north cascades • NOCA** 29

## CALIFORNIA

yosemite • YOSE 33
joshua tree • JOTR 41
death valley • DEVA 49
sequoia • SEKI 55
kings canyon • SEKI 61
lassen volcanic • LAVO 65
redwood • REDW 71
channel islands • CHIS 77
pinnacles • PINN 81

## INTERMOUNTAIN WEST

grand canyon • GRCA 85
rocky mountain • ROMO 93
zion • ZION 101
yellowstone • YELL 109
grand teton • GRTE 117
glacier • GLAC 125
bryce canyon • BRCA 133
arches • ARCH 141
capitol reef • CARE 147
saguaro • SAGU 153
canyonlands • CANY 159
petrified forest • PEFO 163
mesa verde • MEVE 169
carlsbad caverns • CAVE 173
great sand dunes • GRSA 177
big bend • BIBE 181
black canyon of the gunnison • BLCA 185
guadalupe mountains • GUMO 189
great basin • GRBA 193

## MIDWEST


cuyahoga valley • CUVA 197

indiana dunes • INDU 201

hot springs • HOSP 207

badlands • BADL 211

theodore roosevelt • THRO 217

wind cave • WICA 223

voyageurs • VOYA 229

isle royale • ISRO 233


## EAST


great smoky mountains • GRSM 237

acadia • ACAD 245

shenandoah • SHEN 251

everglades • EVER 257

mammoth cave • MACA 265

biscayne • BISC 271

congaree • CONG 275


## ISLANDS


hawai'i volcanoes • HAVO 279

haleakalā • HALE 287

virgin islands • VIIS 293

dry tortugas • DRTO 297

national park of american samoa • NPSA 301


## ALASKA


glacier bay • GLBA 305

denali • DENA 309

kenai fjords • KEFJ 315

wrangell-st. elias • WRST 319

katmai • KATM 323

kobuk valley • KOVA 327

lake clark • LACL 328

gates of the arctic • GAAR 329

good books 334

about the authors 335

about the illustrator 335

acknowledgments 336

index 338

**For Jeb, who helped raise me to love books, and who would have loved this one especially.**

–EMILY

**For CJ, Dawn, and Amy, who are the best siblings out there.**

–MAYGEN

# introduction

We know this *looks* like a book, but our collection of pages is actually more like a secret decoder ring or a pair of X-ray glasses because it will help you see some of the most iconic landscapes in the United States in a whole new way. Whether you're traveling through the national parks by car, bicycle, boat, or foot, or even in your imagination, this is an opportunity to unlock the scientific stories behind the scenery.

This guidebook will teach you to spot the extraterrestrial-like organisms lurking in Yellowstone, the spiky teddy bear clones in Joshua Tree, the slick snails of Acadia—and more! Contained here are true stories about plants, rocks, animals, bodies of water, and the night sky that you aren't likely to find anywhere else than in these parks. We've steered away from people-centric history and from big, obvious questions (like, How did the Grand Canyon form?) in favor of more fascinating, offbeat questions (like, How are strange ocean animals that look like plants connected to the rocks that make up the Grand Canyon?). This is an invitation to be inquisitive and pay attention to the small details that bring the big picture into view.

We had a blast writing this book and hope our work sets you off on a question-asking frenzy of your own. Go forth and get curious!

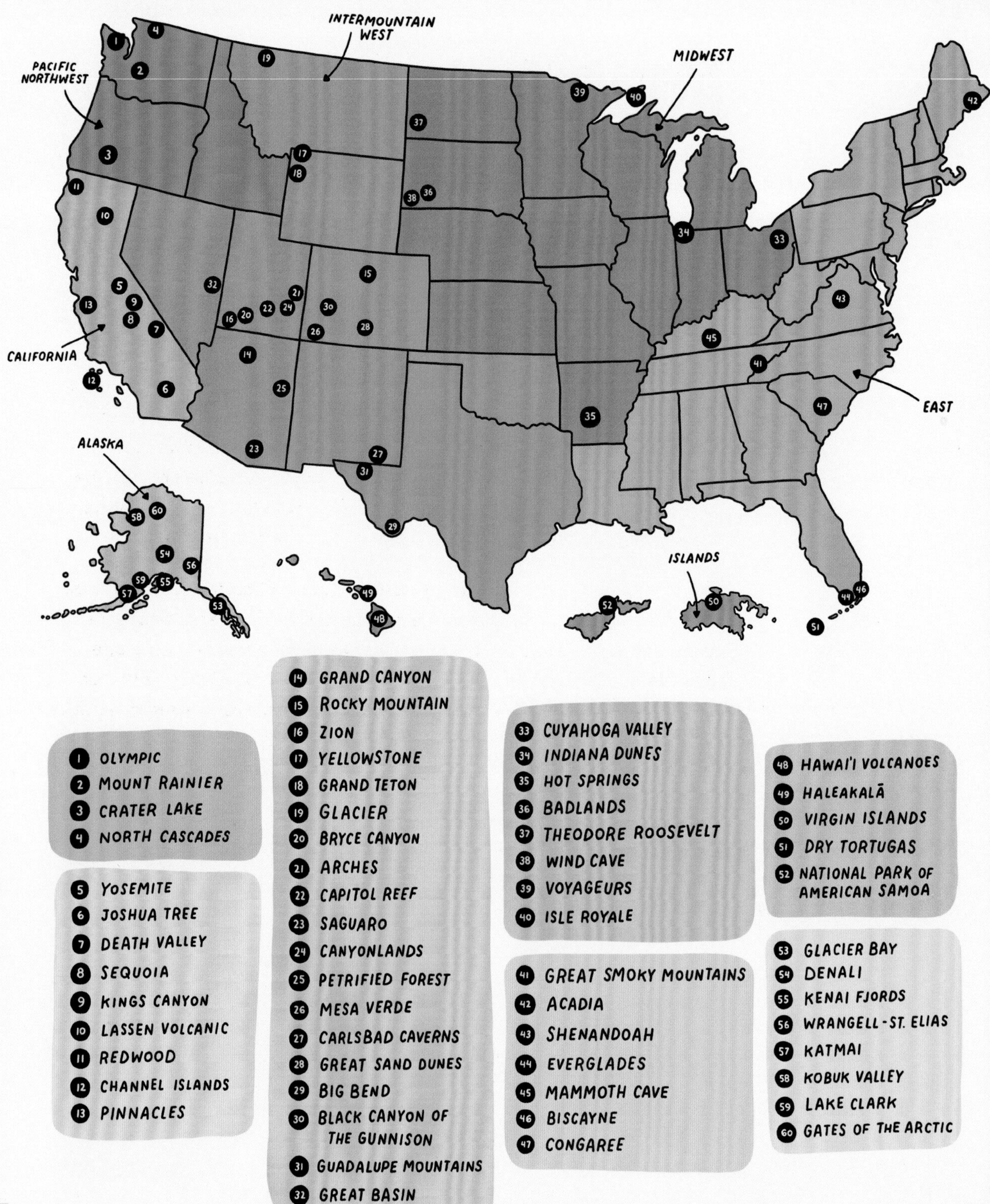
INTERMOUNTAIN WEST
PACIFIC NORTHWEST
MIDWEST
CALIFORNIA
EAST
ALASKA
ISLANDS
1 OLYMPIC
2 MOUNT RAINIER
3 CRATER LAKE
4 NORTH CASCADES
5 YOSEMITE
6 JOSHUA TREE
7 DEATH VALLEY
8 SEQUOIA
9 KINGS CANYON
10 LASSEN VOLCANIC
11 REDWOOD
12 CHANNEL ISLANDS
13 PINNACLES
14 GRAND CANYON
15 ROCKY MOUNTAIN
16 ZION
17 YELLOWSTONE
18 GRAND TETON
19 GLACIER
20 BRYCE CANYON
21 ARCHES
22 CAPITOL REEF
23 SAGUARO
24 CANYONLANDS
25 PETRIFIED FOREST
26 MESA VERDE
27 CARLSBAD CAVERNS
28 GREAT SAND DUNES
29 BIG BEND
30 BLACK CANYON OF THE GUNNISON
31 GUADALUPE MOUNTAINS
32 GREAT BASIN
33 CUYAHOGA VALLEY
34 INDIANA DUNES
35 HOT SPRINGS
36 BADLANDS
37 THEODORE ROOSEVELT
38 WIND CAVE
39 VOYAGEURS
40 ISLE ROYALE
41 GREAT SMOKY MOUNTAINS
42 ACADIA
43 SHENANDOAH
44 EVERGLADES
45 MAMMOTH CAVE
46 BISCAYNE
47 CONGAREE
48 HAWAI'I VOLCANOES
49 HALEAKALĀ
50 VIRGIN ISLANDS
51 DRY TORTUGAS
52 NATIONAL PARK OF AMERICAN SAMOA
53 GLACIER BAY
54 DENALI
55 KENAI FJORDS
56 WRANGELL-ST. ELIAS
57 KATMAI
58 KOBUK VALLEY
59 LAKE CLARK
60 GATES OF THE ARCTIC

# how to use this book

Taking a cue from the National Park Service (NPS), we have organized the sixty national parks featured in this book by region: Pacific Northwest, California, Intermountain West, Midwest, East, Islands, and Alaska. You'll find some handy color-coded page edges to help you navigate. Within each region, parks are listed in descending order based on annual visitation at the time of writing.

At the beginning of each park's profile, you'll find a spot to get your National Park Passport Stamp from a visitor center and a page featuring some park statistics, including the park's four-letter code. These are the same codes the Park Service uses—real insider info! Known as Alpha Codes, the formula for them goes like this: If a park name is one word (Yosemite), the code is the first four letters (YOSE). For names that are two or more words (Crater Lake), the code is the first two letters of the first two words (CRLA). As with any system, there are exceptions, including two sister parks that share the same code.

Within each profile, we've highlighted one to three particularly notable stories about wildlife, geology, botany, and a few wildcards. All of the sights, sounds, smells, and experiences we talk about can be found in the main, developed areas of the parks. Many storylines provide illustrations to give you a better sense of what to look for. We've also included some cross references in the text, but use the index to find additional information about subjects that might cross several parks.

As you head out to explore, two pieces of basic gear will make all the difference: sturdy binoculars are great for daytime views and double as a telescope at night; and a loupe, hand lens, or small magnifying glass will help you see small details on plants and rocks.

At the end of each section, look for this binoculars icon for more information on how to see and experience the park features we've discussed.

# why we wrote this book

As best friends and avid fans of the national parks, we noticed that there are far more science stories happening in each park than could ever be interpreted by staff and scientists—the land is just too remarkable and vast. Even though research about some of the "lesser known" science is happening, there is only so much that can be curated for public consumption. What's a park-loving science-nerd to do?

We decided to take on that problem. We did endless hours of research to find compelling features, talked to the amazing people studying the science of how those features work, and tried to break what we call the artificial barrier between scientists and the general public. We especially wanted to hear from women doing this research in order to amplify their voices in the parks and in science. We did our best to curate content that will inspire your own curiosity while delighting you along the way.

# land recognition

The land that now makes up the United States of America and its territories has been occupied, cared for, studied, and deeply understood by Native and Indigenous peoples for tens of thousands of years. They are the original storytellers and record keepers of these places, and they continue to maintain ancient and unending ties to the land. Many modern Native groups still live close to public lands of many kinds, including national parks, and they have fought hard to be able to continue their relationships with these spaces that are now open to all.

Out of respect for the people who are the original stewards of these amazing places, we encourage you to take time during your trip planning or armchair perusing to learn about the historic and modern Native groups associated with the locations you will visit. When you're in the parks, take a moment to explore nearby tribal tourism or cultural centers and consider staying in a Native-owned hotel, finding a Native-led guided tour, or visiting a community market to support local vendors.

We are not the first people to find these landscapes awe-inspiring and worthy of celebration. Let's honor those who cared for them first and know them best by learning about their cultures and respecting their homelands.

# safety in the parks

When you're out and about in the parks, there are many safety issues to take into consideration. This list is not comprehensive, but here are some general safety tips to keep in mind as you explore. Be sure to always use common sense, and to visit the visitor center when you arrive at a park to ensure you know its particular safety policies—each park is unique!

## DRIVING

- Drive safely and respectfully.
- Observe the speed limit, even if it seems slow—enjoy the scenery!
- Pullouts are great for letting cars pass and for viewing wildlife—DO NOT stop in the middle of the road unless wildlife is crossing.

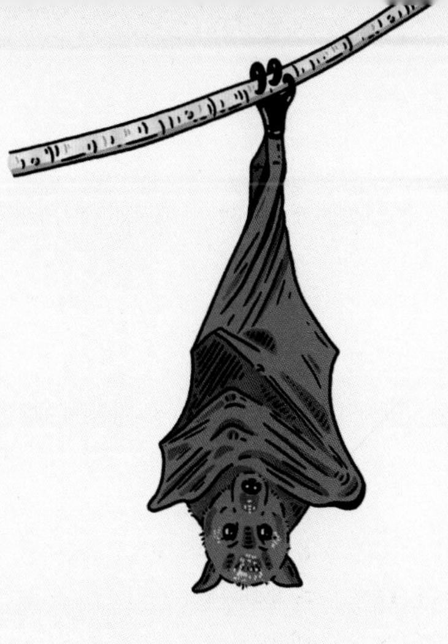

## HIKING

- Know your hiking ability and stay within it. Research the trail before you go.
- Stay on established trails—they're there to protect you and the park.
- Dehydration is always a risk—drink plenty of water and carry at least a couple liters with you.
- Plan hikes according to the weather forecast, but know that weather can change quickly.
- Dress appropriately for the weather and plan for potential weather changes. Layers are always a good idea.
- Make sure someone knows where you're going and when you'll be back.
- Be very cautious around any body of water—drowning is the leading cause of death in the parks.
- Leave no trace—pack out everything with you, including food waste, or dispose of waste in an available garbage can.

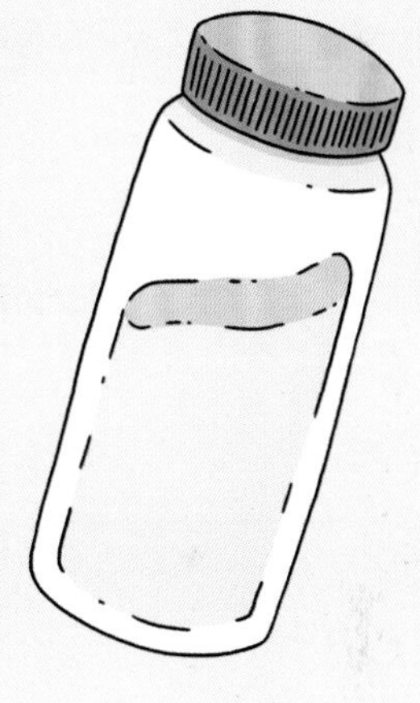

## PROTECTING LAND AND WILDLIFE

- Keep wildlife wild. Never feed, touch, or otherwise engage with an animal. If you're close enough to affect any animal's behavior, you're too close. Typically, staying 100 yards away from any animal is the best course of action.
- Take only pictures—since this is federally protected land, removing anything—from rocks and fossils to flowers and pinecones—is a crime. It's important that this land and the items on it are preserved for future generations to enjoy.

pacific northwest

california

intermountain west

midwest

east

islands

# olympic

STATE
**WASHINGTON**

ANNUAL VISITATION
**3.1 MILLION**

YEAR ESTABLISHED
**1938**

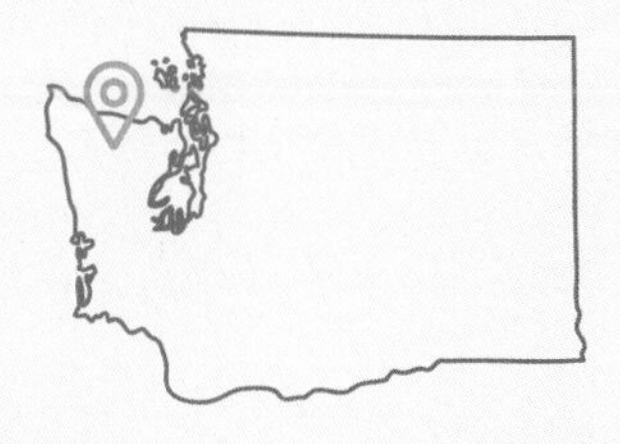

## an isolated peninsula with rain forests & rocky beaches

| | |
|---|---|
| SUPERLATIVE | Mount Olympus is the wettest place in the lower 48 |
| CROWD-PLEASING HIKES | Rialto Beach to Hole-in-the-Wall (easy); Hurricane Hill (moderate); Glacier Meadows (strenuous) |
| NOTABLE ANIMALS | Olympic marmot (*Marmota olympus*); humpback whales (*Megaptera novaeangliae*); banana slugs (*Ariolimax columbianus*; see page 23) |
| COMMON PLANTS | Sitka spruce (*Picea sitchensis*); bull kelp (*Nereocystis luetkeana*); spreading phlox (*Phlox diffusa*) |
| ICONIC EXPERIENCE | Hiking the Hall of Mosses trail |
| IT'S WORTH NOTING | The nearby town of Port Angeles is home to an eccentric and charming community radio station. Tune your dial near the park to KSQM 91.5 FM. |

**WHAT IS OLYMPIC?** The Olympic Peninsula is an isolated chunk of land in the far northwest corner of the contiguous United States. It's *alllll* the way up there, surrounded by water on three sides, covered in towering peaks, receding glaciers, and rushing rivers. Basking in the heart of that isolation is OLYM—largely an immense wilderness almost entirely without roads or human structures. Some of the richest forests in the country are fed by the rivers that radiate from Mount Olympus and by blankets of coastal fog. The cold waters of the Pacific Ocean support an incredible cast of critters along a wild, rocky shore begging to be explored.

# BOTANY

## EPIC EPIPHYTES

**THERE'S NO DENYING THAT THE** trees of OLYM's rain forests are mesmerizing. They're big, tall, and covered in green. Epiphytes, or plants that grow on plants, are a hallmark of rain forests, and OLYM is no exception. These mosses, lichens, liverworts, and ferns grow, hang, drip, and extravagantly cover tree trunks and branches without harming their hosts at all. Epiphytes draw everything they need from dripping water and nutrients blowing in the wind. More than one hundred thirty species of epiphytes have been identified in OLYM's rain forests so far. Though they are on virtually every tree, look for especially grand displays on maple trees.

Bigleaf maples (*Acer macrophyllum*), named obviously for their leaves, are almost unrecognizable under their epiphyte finery. Even if you can't make out the distinctive maple leaf shape, these relatively more squat trees stand out among the taller, slimmer conifers of the canopy. The aerial gardens these trees have cultivated are truly massive; one study found that the epiphytes on a bigleaf maple weighed up to four times as much as the tree's actual foliage.

As you approach a bigleaf, notice the garden of hanging mosses dangling from its branches. At the base of those gardens is a moss mat up to 10 inches thick. That's nine of these books stacked on top of each other! Keep an eye out for the long, hanging tendrils of Oregon spikemoss (*Selaginella oregana*). This fernlike moss easily has the most metal name in the forest and shows off its prowess by growing up to 6 feet in length. Slightly higher in the canopy are impressive ferns, called licorice ferns (*Polypodium glycyrrhiza*); they do indeed have a sweet licorice taste. Look for them on local restaurant menus in the spring.

On lower branches and trunks you'll notice the tiny, feathery tendrils of shiny, green cattail moss. You might see tiny drops of water on the end of the tendrils. You can also look for flaky green plates that resemble lettuce or cabbage. Called

## DRAPED IN FINERY

Bigleaf maples host some of the most diverse epiphyte communities in OLYM's forests.

**Oregon spikemoss (*Selaginella oregana*)** These mosslike green streamers are covered in tiny pointed leaves. As the streamers dry, they curl up and turn brown.

**Licorice fern (*Polypodium glycyrrhiza*)** Single, unbranching fern fronds growing directly out of the moss mat are licorice ferns. Stems are light green or straw colored, and leaflets grow in staggered ranks on opposite sides of the stem.

**Bigleaf maple leaf (*Acer macrophyllum*)**

**Oregon lungwort (*Lobaria oregana*)** This flat-lying green lichen is covered in ridges and has frilly edges. Red to brown spore-bearing structures are often visible. One of the most common lichens in old-growth forests in the Pacific Northwest.

Oregon lungwort (*Lobaria oregana*), this helpful lichen has the ability to grab nitrogen right from the dang air. The nutrients it provides are a key part of keeping this rain forest healthy and well-fed.

It seems remarkable that all these plants (and their combined weight) don't harm the trees, but epiphytes are upstanding members of the community. In fact, recent research has revealed that bigleaf maples sometimes send roots (identical to their underground roots) into the plant growth on their limbs to tap into the water and nutrients of their epiphyte pals. Who wouldn't want to be friends with epiphytes? They're always around, most of them are soft and cuddly, and they even make their own snacks to share with everybody.

**The west-facing Quinault, Queets, Bogachiel, and Hoh River Valleys all host rain forests. Short, flat Maple Glade Rain Forest Trail in the Quinault is a great, less-visited alternative to the busy Hoh. The Queets and the Bogachiel offer more secluded and wild scenery, with some very challenging hikes. Be prepared for rapidly changing weather conditions in these coastal forests.**

**TIDE POOLS FOUND IN THE** Pacific Northwest (PNW) boast fantastic creatures. Brightly colored sea stars cling to solid surfaces, snails scoot casually along the rocks, and anemones hang out waiting to use their tentacles to sting and trap their prey. But as you're walking around this overflow of ocean life, you'll notice some other humble players attached to rocks *everywhere*. Barnacles keep this ecosystem alive, and they have some pretty cool attributes worth a closer look.

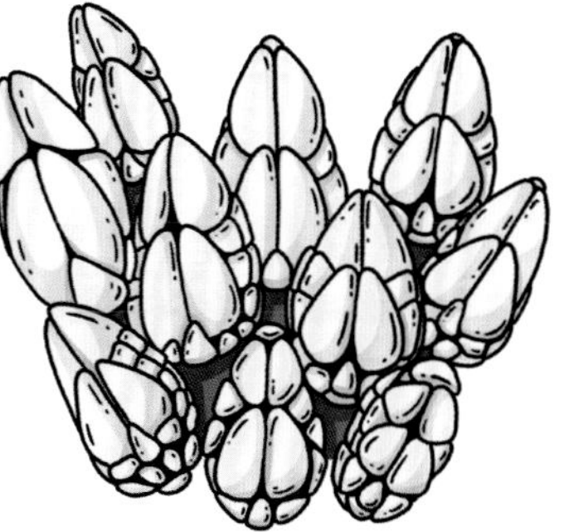

Gooseneck barnacle
(*Pollicipes polymerus*)

Be on the lookout for two types of barnacles on your tide-pool adventure. There are acorn barnacles (*Balanus glandula*), your "typical" barnacle, living among the rocks. These little guys usually range from the size of a pea to the size of a quarter (although giant acorn barnacles can be the size of a human fist!). They are usually gray and live in clusters of tens to hundreds to thousands. Also living on the rocks are gooseneck barnacles (*Pollicipes polymerus*), which have 3- to 4-inch-long fleshy necks and are more of a conical shape. If you're lucky, you might find larger goosenecks (9 to 10 inches long!) attached to logs washed up on shore.

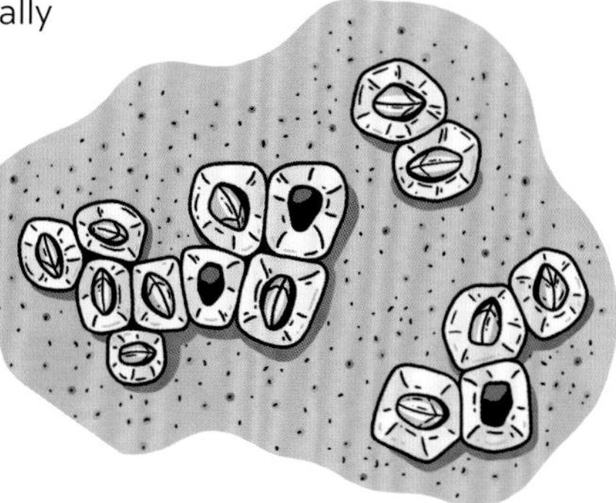

Acorn barnacle
(*Balanus glandula*)

Barnacles are crustaceans, and they begin life as free-floating creatures in open water, searching for a place to call home. By the time we see them, they've glued their heads to a surface

WILDLIFE

## BARNACLES AND THE GLUE THAT KEEPS THEM STICKING

Giant barnacles, thatched barnacles, and pelagic goose barnacles are three other barnacles you might find while you're out tide pooling!

and built hard shells around themselves, where they'll remain until they die. Since they can't move, they feed and mate with help from a trapdoor-like opening at the top of their shells. When barnacles are covered with water, their door opens and they use feathery legs to catch microscopic plankton. At low tide, when we see barnacles, the doors are closed to keep the creatures from drying out.

While they may seem like an insignificant part of the inter-tidal zone, barnacles are actually crucial to the success of the ecosystem. Small areas between the barnacles are often home to other creatures that need to stay moist at low tide. Barnacles also provide food to other intertidal creatures, like carnivorous snails. The snails drill into the barnacles' shells and digest their soft bodies. Barnacles' most crucial role may be restoring nutrients to the water after digesting their food, allowing life to continue to thrive where land meets sea.

Barnacles produce a glue that can stick to pretty much anything, including the hulls of ships and even the bodies of whales, which is pretty remarkable—and for a long time was not understood. In 2009, scientists finally solved the mystery of how barnacles' glue can cling so firmly to under-water surfaces. They found that the glue uses an enzyme genetically similar to the one that allows red blood cells to clot to make scabs. This enzyme enables a barnacle's glue to cure in water. So before you try to rip that little barnacle off just like a scab on your elbow, take a moment to think about how you're somehow evolutionarily related. Then leave that barnacle alone!

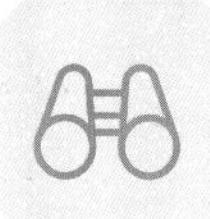

**Visit the coastal areas of the park, like Rialto Beach. Be sure to check a tide chart so that you're there when the tide is low. There are a lot of creatures to see, including sea stars, crabs, and anemones, so keep your eyes peeled. Wear good shoes and be mindful where you step; these rocks can be incredibly slick!**

## WILDCARD

## (UN)DO THE DAM THING

**FROM ITS HEADWATERS HIGH IN** the Olympic Mountains, the Elwha River careens down thousands of feet before spilling into the saltwater of the Strait of Juan de Fuca. Most of its 45-mile length is contained within OLYM. Swift elevation change and narrow gorges make the Elwha a powerful force. Eager to harness its power, Euro-Americans built two dams on it in the early 1900s. In the process, the once life-giving Elwha became choked and barren.

Decades of tireless advocacy by many, including the Lower Elwha Klallam Tribe—whose creation place is along the river—brought the dams down in 2014. Now, the Elwha is reclaiming its land and its power, and you can watch it happen.

Almost everything changed when the Elwha was dammed. The dams choked the river's supply of sediment and plant material from upstream, depriving downstream species of important places to live and hide from threats. Piles of plant material along the banks and tangled on rocks and other debris in the water might seem unremarkable to you, but they are a vital component of this river ecosystem.

Crucially, the dams also blocked the migration of salmon from the ocean. Their absence deprived the ecosystem of key nutrients. Now, though, the fish are once again swimming upstream. Some salmon get eaten along the way and some make it upstream and lay eggs (some of which also get eaten). Once the salmon have spawned, they die. Any way you slice that salmon, it's doling out nutrients to predators, scavengers, and even plants as the leftover scraps and animal scat decompose into fertilizer.

Birds are one key indirect salmon beneficiary. Look and listen for gray American dippers (*Cinclus mexicanus*) along the river. They're small, tubby birds with white-feathered, flashing eyelids and a sharp "zeet" call. They eat small aquatic invertebrates that are flourishing thanks to salmon nutrients. In the time since the dams were removed, scientists have found that the dippers are staying in the area longer (rather than migrating out), and some of them are even raising two broods of chicks per year.

The river's return has altered plant life too. If you're lucky and arrive while riverbank lupine (*Lupinus rivularis*) is in bloom, you might just smell it before you see it. These purple to white flowers were absent from the Elwha while the dams were in place, but they are back after former reservoirs were reseeded with native plants. Such revegetation efforts aim to establish woody plants and trees and help native plants take hold, as well as discourage nonnative plants from moving in.

Riverbank lupine fixes nitrogen in the soil—it's a good neighbor to have around.

The impacts of the removal of the Elwha dams will continue to be studied and monitored—nature isn't an on/off switch that can be thought of as broken or healed. But the river, thanks in large part to its human guardians, the Elwha Klallam and the National Park Service (NPS), is healing itself right before our eyes.

**The undammed Elwha is prone to floods that can damage infrastructure. Check in with park staff about current conditions before finalizing your plans. Hiking trails, some quite strenuous, wind through the Elwha Valley. Madison Falls, right on the park boundary, is the shortest and most easily accessible. Overlooks on both sides of the river feature views of the Glines Canyon Dam site.**

STATE
**WASHINGTON**

ANNUAL VISITATION
**1.5 MILLION**

YEAR ESTABLISHED
**1899**

# mount rainier

## a glacier-covered active volcano

| | |
|---|---|
| SUPERLATIVE | The Sunrise Visitor Center is the highest elevation point in Washington accessible by car |
| CROWD-PLEASING HIKES | Silver Falls (easy); Naches Peak Loop Trail (moderate); Skyline Trail (strenuous) |
| NOTABLE ANIMALS | Banana slugs (*Ariolimax columbianus*); hoary marmots (*Marmota caligata*); black-tailed deer (*Odocoileus hemionus hemionus*) |
| COMMON PLANTS | Fireweed (*Chamerion angustifolium*; see page 320); Douglas-fir (*Pseudotsuga menziesii*); western white pine (*Pinus monticola*) |
| ICONIC EXPERIENCE | Seeing a meadow full of wildflowers |
| IT'S WORTH NOTING | It's possible to hike smaller sections of the park's famed 93-mile-long Wonderland Trail. |

**WHAT IS MOUNT RAINIER?** One-third of MORA is the actual Mount Rainier—the most glaciated peak in the lower 48 and one of a number of active volcanoes in the Cascade Range. The other two-thirds of the park boast forests, glacial lakes and streams, and stunning wildflower-filled meadows. Take a trip to Sunrise to soak up fantastic views and head out on day hikes with unforgettable scenery. Scattered through the park are patches of old-growth forest with trees more than a thousand years old. Much younger forests thrive in places that haven't escaped Rainier's recent volcanic eruptions, mudflows, and floods.

# GEOLOGY

## FIRE, ICE, AND MUD

**FIRE HAS FORGED MUCH OF** Mount Rainier. But ice is the real architect of this volcano, and it continues to sculpt it today in ways you might not expect.

The glaciers that now flank Rainier are mere shadows of those that have covered it in the past. At times, ice more than a thousand feet thick trapped and directed lava flows from the volcano. Erupting lava was forced to pool between sheets of ice, so that when the glaciers melted back, fins of cooled lava were left jutting steeply above the surrounding terrain. The top of the ridges today more or less marks the depth of ice during the eruption that created them. Look for them and try to imagine the hiss of lava meeting ice.

Glaciers large enough to shape lava flows no longer exist on Rainier, but that doesn't mean that the glacier-volcano collaboration is over. When there is a sudden release of water from the glaciers, such as after a large rainstorm, cascades of debris-filled water, volcanic ash, and broken rock bulldoze their way down river valleys in the park, obliterating anything in their paths. Known as debris flows, these dynamic events can rampage downslope at 20 to 30 miles an hour, wielding boulders and trees like battering rams. Rushing water, snapping wood, and shattering rocks create a cacophony that some have compared to the sound of a runaway train off its tracks.

Occasionally, factors conspire to create a debris flow of epic proportions. In MORA, debris flows that travel beyond park boundaries are called *lahars*—the Indonesian word for "mudflow." Lahars are usually triggered when volcanic eruptions cause glaciers to melt suddenly, releasing a torrent of water mixed with hot pyroclastic material, or when soft rock near the summit gives way. Volcanic steam vents high on the mountain slowly alter hardened lava into weak clay, making the summit of Rainier especially susceptible to collapse. In the past, whole sides of the volcano have crumbled, sending lahars tearing through river valleys as far as 50 miles downstream from the volcano.

Debris flows can happen at any time but are most common during the heat of late summer and heavy rainstorms in autumn and early winter. If you hear a prolonged rumbling, head to higher ground immediately. Lahars are bigger and require evacuation of entire communities. You cannot outrun a debris flow or lahar.

## A VOLCANO SHAPED BY ICE

Look up and look around as you explore MORA—evidence of the growth of a volcano shaped by glaciers is everywhere!

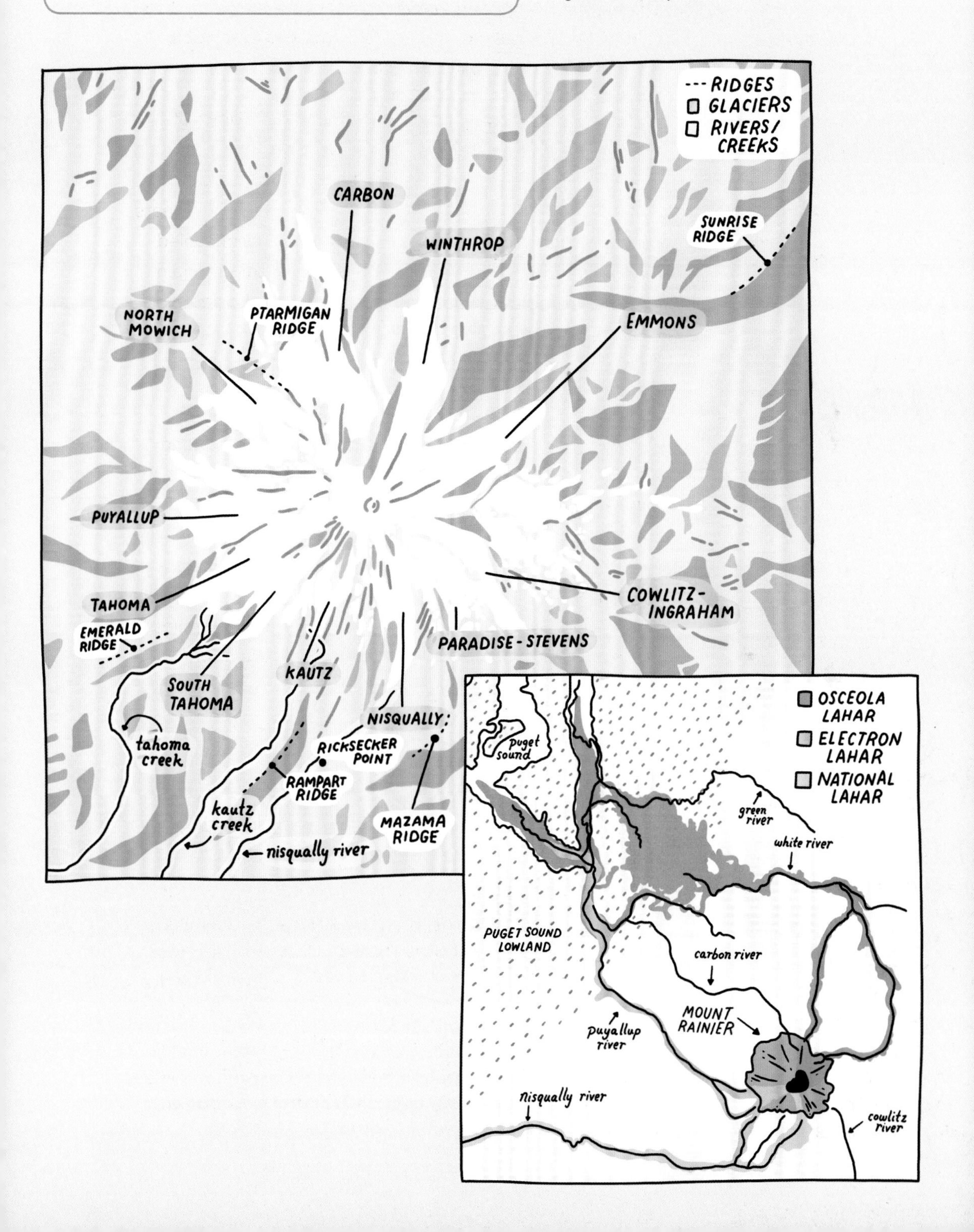

When rocks are smashed and broken open, their minerals react with the air—that's what causes the water in Kautz Creek to look rusty.

The walls of river valleys and the banks of creeks often contain records of past debris flows and lahars. The most obvious signs of past catastrophes are the basketball-or-larger-size boulders suspended in the mud of the banks. On slightly higher ground, you may be able to spot dead standing trees, called snags. These trees were partially buried by debris flows and slowly died in place.

Keep an eye out for large boulders scattered seemingly at random around the park. These boulders were dropped in their present locations after the more liquid part of a debris flow drained away, leaving them like marshmallows stranded on a table after a hot chocolate spill. Boulders colored slightly yellow or brownish come from closer to the mountain's summit, where they were altered by hot volcanic gases.

The landscape in MORA, much like Mount Rainier itself, is a storyteller that delights in rehashing its antics to anyone keen enough to listen. What will the volcano and its icy mantle tell you?

**The high ridges that radiate out from Mount Rainier are mostly cooled lava fins. They are easily visible on the volcano's southwestern side. At the Paradise Jackson Visitor Center, you are standing on top of a stack of chilled lava flows. Kautz Creek, Tahoma Creek, and the Nisqually River are all good places to look for evidence of past flows. Mudflow boulders are scattered throughout the park, with some especially large and colorful examples near the Paradise Jackson Visitor Center.**

## BOTANY

### THE LOGS THAT KEEP ON GIVING

**DECOMPOSITION IS A MAGICAL PROCESS** that most of us understand in theory, but with which we generally have little interaction. We know that it happens, we know it's helpful, and we know that living things will decompose in some capacity when they die. One of the beautiful gifts that forests in the Pacific Northwest give us is the opportunity to see decomposition up close.

More than half of MORA is forest. Head out into those woods, especially in the lower elevations of the park, and notice the cluttered mess of fallen logs, mosses, snags, and other debris. This massive mess of material is critical for the entire forest ecosystem to function. In the park, up to 150 tons of debris per acre has been measured. The stars of the show, the nurse logs, give an excellent view of death directly promoting new life. Nurse logs are fallen trees that foster new growth. Common in the PNW, they are typically fallen Sitka spruce, Douglas-fir, and western hemlock.

Newly fallen nurse logs often don't have much more than a few mosses growing on them. They look pretty similar to fallen logs you've seen in other forests. On slightly older nurse logs, look for small ecosystems of plants, young trees, and animals. If you keep your eyes peeled, you will be able to spot nurse logs that are almost completely decayed. Look for a tree or row of trees that appear to be standing on stilts.

After a massive tree falls, fungi and bacteria, along with insects and other invertebrates, begin the process of decomposition. Mosses take hold, allowing seedlings from other plants to establish their root systems

A tree growing from an aged nurse log.

In the PNW, it takes 500 to 600 years for a large fallen log to become 90-percent decomposed.

securely in the moss. Without nurse logs, these seedlings could become easy snacks for forest wildlife such as elk, or they could simply blow away. Putting down roots is just one part of the battle, though. Seeds need light to photosynthesize, and the shadowy forest floor isn't exactly sunny and bright. When you come across a nurse log, look up and around. Notice that the fallen tree created openings for light to shine through, helping life start over on its remains.

As time passes, fungi continue to decompose the log, turning it into super-nourishing sustenance for the growing seedlings. Decades pass while the log decays and new life flourishes. The log doesn't just provide nutrients as it decomposes; look for small nooks and crannies in the log—you might see snails, banana slugs, spiders, and even vertebrates, like squirrels, making themselves at home.

Consider this during your time scoping out nurse logs of any age: Most of a living tree's trunk is dead tissue—once a fallen tree becomes a nurse log, it actually fosters much more life than it did when it was alive.

**Take a hike in an old-growth forest in the park. Grove of the Patriarchs Trail is one of the easiest in MORA, and if you keep scanning the forest floor, you should see nurse logs in various states of decomposition.**

# WILDLIFE

## WHAT NOW, MY SLUG?

**MORA IS HOST TO A** plethora of animal species. Birds, mammals, reptiles, amphibians—the park has all of these. Yet you might not be thinking about the most populous and diverse group of creatures in the park: the invertebrates! They're literally everywhere, from the damp forest floor all the way up to the highest peaks (over 14,000 feet). According to the National Park Service (NPS), these spineless critters likely make up about 85 percent of the park's animal biomass. *Wow.*

MORA's banana slugs (*Ariolimax columbianus*) are the only slug native to the PNW. Sometimes they're yellow, like bananas, but they can also be dark green or brown. They're also pretty large, growing to between 6 and 10 inches in length. Even though they're abundant in the park, they're still a treat to come across. First, head out for a hike in a wet, forested area. Keep scanning the ground—you might find a slime trail leading you right to a banana slug. Look for small logs, tree roots, or any other super-moist area—if you (gently) lift up logs, you might find a slug lurking underneath. Banana slugs can also use their slime to create a slime cord, which they use to gently drop down from tree branches—keep your eyes peeled!

Banana slug
(*Ariolimax columbianus*)

Look at the forest floor for mushrooms too—banana slugs love a good 'shroom. They're not picky, though. Key members of the ecosystem, slugs help put nutrients back into the soil after they digest their food. These slugs will snack on anything from fallen leaves to lichens to fruits to poison oak (thanks, slugs!), and they'll even feed on the carcasses of dead animals. If it can be eaten, they'll eat it.

Slugs are hermaphrodites, which means they have both male and female sex organs. However, typically they don't self-fertilize. Instead, they examine a potential mate's slime, then lick each other's right sides, where the genital opening is. This goes on for several hours, and then they connect for several more hours, when fertilization takes place. Occasionally, after all that time, the slugs can get stuck, leading to apophallation—the technical term for slugs biting off a penis to get free.

After all this, each fertilized slug lays eggs in a moist, safe spot and moves on with its life. As you're out there slug-hunting, remember that slugs are animals just like all the other wildlife in the park. Slugs are pretty clear when they feel threatened—they shorten their bodies and hide their heads. If they do that, back away. Instead of getting too close, snap a quick photo, enjoy that amazingly long slug body, and then tell everyone you know that banana slugs some of the baddest beasts out there.

**Keep your eyes peeled while you're in woody areas of the park. Banana slugs can be found anywhere, including on boardwalks! Remember to keep your hands off all wildlife you find in the park.**

STATE
**OREGON**

ANNUAL VISITATION
**721,000**

YEAR ESTABLISHED
**1902**

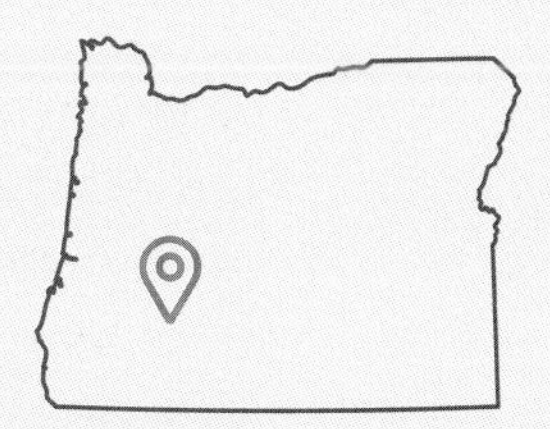

# crater lake

## the beautiful result of a devastating eruption

| | |
|---|---|
| **SUPERLATIVE** | Crater Lake is the deepest lake in North America |
| **CROWD-PLEASING HIKES** | Sun Notch (easy); Watchman Peak Trail (moderate); Garfield Peak (strenuous) |
| **NOTABLE ANIMALS** | Black-backed woodpecker (*Picoides arcticus*); Mazama newt (*Taricha graulosa mazamae*); golden-mantled ground squirrel (*Spermophilus lateralis*) |
| **COMMON PLANTS** | Ponderosa pine (*Pinus ponderosa*; see page 88); lodgepole pine (*Pinus contorta*; see page 113); woolly Indian paintbrush (*Castilleja foliolosa*) |
| **ICONIC EXPERIENCE** | Jumping into the water at Cleetwood Cove |
| **IT'S WORTH NOTING** | If you can peel yourself away from the water, the Pinnacles and the Pumice Desert show off a different side of the volcano. |

**WHAT IS CRATER LAKE?** This "crater" is actually a caldera—the remnants of a volcano-shattering eruption a mere 7,700 years ago. Mount Mazama, one of the Cascade volcanoes, stood about 12,000 feet tall before the eruption that collapsed it into what you see today. Rain and snow have steadily filled the caldera over the ensuing centuries, producing a lake of astounding clarity and purity that is an impossible shade of blue. The volcano didn't retire after the caldera-forming eruption—subsequent activity built Wizard Island in the lake, and heat continues to gurgle beneath the water's surface. Many parks offer volcano views, but none are quite like CRLA!

# GEOLOGY

## A WATERY VANISHING ACT

Glaciers also carved the soft U-shaped notches you can see in the caldera rim.

**CRATER LAKE IS DEFINITELY A** lake like no other. Inside a giant caldera, it isn't fed by rivers or streams. Instead, all of its water comes from precipitation—mostly in the form of snow during the long, frigid winters. It also breaks one of the cardinal rules of lakes in general: It doesn't completely fill its basin. Instead of a marshy shoreline, it is ringed by hundreds of feet of bare rock.

On average, Crater Lake receives about 80 inches of rain and snow annually. That's pretty much the only water that's added to the lake every year. The chilly climate means the lake loses only about 30 inches of water a year to evaporation. That's a difference of about 50 inches that goes in and doesn't seem to come out. But the lake doesn't rise that much each year. Its water level is remarkably steady—it fluctuates when precipitation patterns change, but only within a narrow range. Something is funky in the water accounting department here.

Where is the water going? It's seeping through the caldera wall! Most of the caldera is made of cooled lava—kind of what you'd expect from a volcano. But not all of it. Look for Palisade Point on the northeastern section of the rim. Just below that, and behind the rock debris (called *talus*; see page 255), is a layer of crumbled and broken rocks that were left behind by the retreat of a glacier on Mount Mazama several hundred thousand years ago and subsequently incorporated into the volcano's flanks. Much looser and more permeable than its igneous neighbors, the glacial rubble lets water seep through the northeast side of the caldera. The rubble pile extends from below the cliffs to 140 feet underwater—it acts just like the overflow drain on your bathtub and keeps Crater Lake from fully filling its basin.

Every hour more than 2 million gallons of water leaks through the caldera wall. That's more than enough to fill three Olympic-size swimming pools every single hour of the day! The seepage goes on all day, every day, which is to say a *lot* of water is working its way through the caldera wall at any given moment. It still isn't enough to offset all of the precipitation that falls here, so the lake's levels remain constant.

## UNDER THE DEEP BLUE

Maps of Crater Lake's floor reveal a complex world. Glacial rubble near Palisade Point lets water seep through the caldera wall, keeping the water levels remarkably constant.

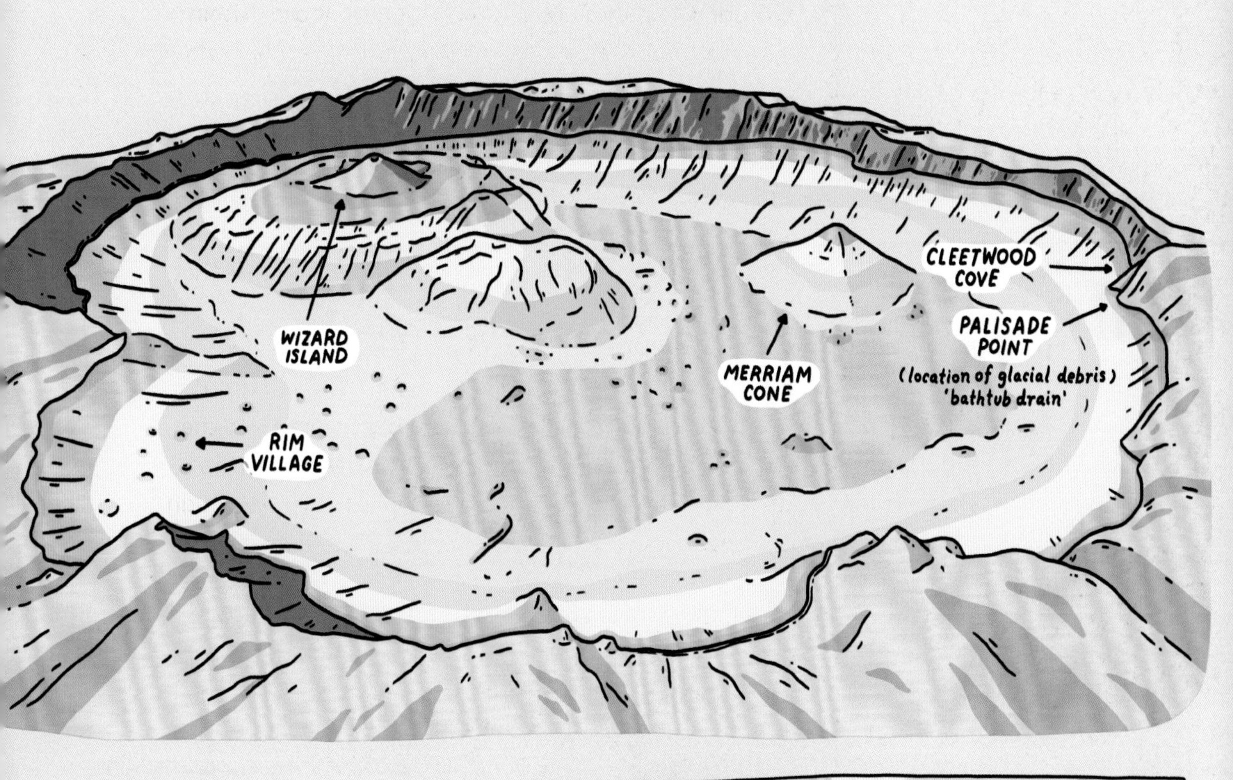


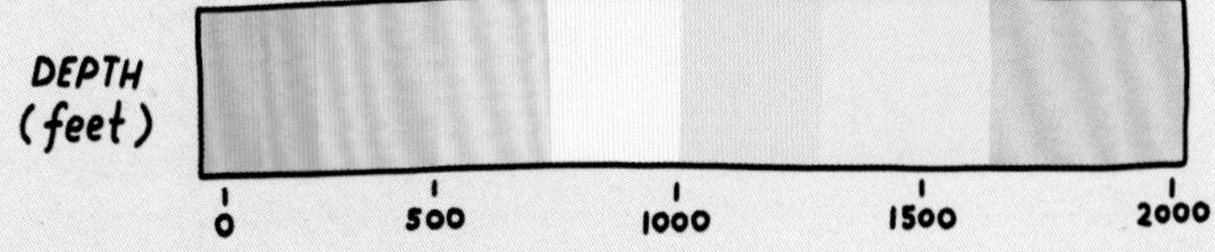

You might expect to now be told where Crater Lake's errant water goes. It's a reasonable assumption—but the crazy part is, no one knows where this water goes! And don't think people haven't looked, either. Water from the lake has a pretty distinct chemical signature because hydrothermal vents on the caldera floor inject the water with telltale amounts of elements such as lithium and boron. Scientists have tested springs, creeks, streams, pools, and puddles all over this area and have yet to find the wily water. Two springs, Oasis and Crater, quite possibly have some lake water in them, but it's no more than 7 percent of their total water and a very insignificant amount from the lake.

As you look out at the impossibly blue water, Crater Lake feels a little mysterious. People witnessed its formation, scientists have probed its depths, but, at least for now, where its waters go is a secret only Crater Lake knows.

**From the rim or from a boat, views of the water are everywhere in the park.**

STATE
**WASHINGTON**

ANNUAL VISITATION
**30,000**

YEAR ESTABLISHED
**1968**

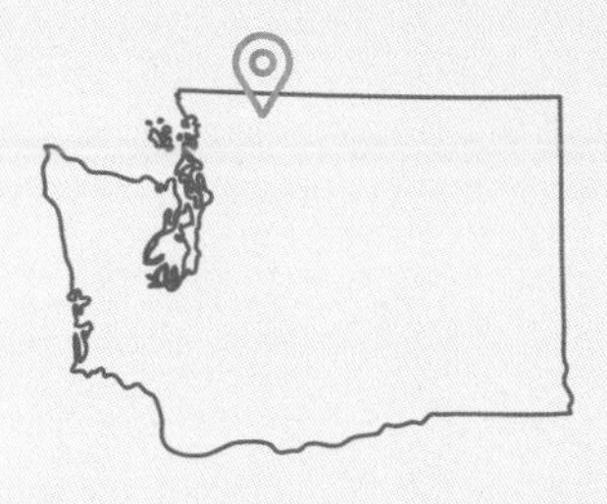

# north cascades

## rugged peaks festooned with glaciers & waterfalls

| | |
|---|---|
| SUPERLATIVE | NOCA contains the most glaciers of any park outside Alaska |
| CROWD-PLEASING HIKES | Agnes Gorge Trail (easy); Fourth of July Pass (moderate); Thornton Lakes Trail (strenuous) |
| NOTABLE ANIMALS | Pika (*Ochotona princeps*); harlequin duck (*Histrionicus histrionicus*); arctic fritillary butterfly (*Boloria chariclea*) |
| COMMON PLANTS | Sagebrush (*Artemisia* spp.; see page 123); sword fern (*Polystichum munitum*); western hemlock (*Tsuga heterophylla*) |
| ICONIC EXPERIENCE | Spending the night at a boat-in campsite on Ross Lake |
| IT'S WORTH NOTING | You can hike a (very) short portion of the 2,600-mile-long Pacific Crest Trail along Cutthroat Pass. |

**WHAT ARE THE NORTH CASCADES?** The Cascade Range is an arc-shaped band of explosive volcanic centers (like Mount Rainier and Mount Mazama) that runs from Northern California to British Columbia. It is the Pacific Northwest's contribution to the Pacific Ring of Fire. The range is made up of plucky, extremely active mountains. NOCA preserves one portion of this young, rugged landscape. Managed alongside Ross Lake and Lake Chelan National Recreation Areas, the park itself is a true wilderness crisscrossed by rough and rewarding hiking trails. Bright-blue glacial lakes and innumerable waterfalls—source of the mountain range's name—make hiking, paddling, camping, and climbing here totally unforgettable.

## BOTANY

### GROOVING IN OLD-GROWTH GROVES

**MANY PEOPLE HABITUALLY ASSOCIATE THE** Pacific Northwest with lush forests of ferns and huge trees. In some ways that association is left over from a time when Europeans were first pushing west. Astounded by the sizes and hardiness of the trees in the PNW, many settlers began clear-cutting forests (taking neither measurements nor prisoners, as some big-tree fans have observed) to turn all those trees into commercial goods.

Luckily for us, just like back East in the Smoky Mountains (see page 237), some Western areas escaped the greedy settlers' saws and survived to whisper their stories to those sharp enough to pay attention. NOCA is one place where it's still possible to wend your way through ancient groves of trees, especially in the lower elevations. These ecologically sacred places are often called ancient, or old-growth, forests—a somewhat loose classification that means the trees show little to no sign of Euro-American disturbance and are often hundreds of years old.

In the absence of flashing neon signs advertising "See Old Growth Here," there are a few ways you can recognize these special forests. Huge trees are a good tip-off, but true old-growth forests are complex and layered. Look for multitiered canopies made up of trees of different heights and species. Look for a mix of ages too. On the forest floor, you might see seedlings interspersed with dead trees and decomposing logs. In the canopy, you'll see living trees in various stages of maturity mixed in with standing dead trees (known as snags).

You can reliably spot three species of tree in NOCA's old-growth groves. Douglas-fir (*Pseudotsuga menziesii*) are among the most abundant and tallest trees. They can live up to a thousand years in the right circumstances. Look at the bark for a quick and easy ID: Mature trees have dark brown to gray bark broken into vertical plates that are streaked with grooves up to 8 inches deep. Branches on old-growth Doug-firs (an adorable pet name) often don't start until relatively high up on the trunk. Check out the ground beneath the tree for fallen cones—Doug-firs' cones have a unique three-pronged bract, resembling the rear legs and tail of a mouse, sticking out between the rounded scales. Bracts are modified leaf structures that pop up in a lot of plants, including on other coniferous cones, but none are quite like the bract on a Doug-fir cone.

A Douglas-fir cone

Western hemlocks (*Tsuga heterophylla*) thrive in the shade of taller Douglas-firs. Look for their characteristic tops that curve and droop to one side. (Most conifer tops stand straight up.) Their cones, which are usually less than an inch long, are made up of thin, rounded plates. Hemlocks grow quickly but often bide their time in the forest understory for centuries before reaching the canopy. Their main claim to fame is their role as nurse logs (see page 21). Hemlocks that die and fall over become rooting grounds for a new generation of trees. Keep your eyes peeled for colonnades of trees growing in a straight line along what was once the trunk of a fallen hemlock. Sometimes, if the original log was quite massive, when it eventually rots away the younger trees' roots look like stilts holding the trees above the forest floor.

Scanning the forest, you may also notice the stringy red-and-brown bark of western red cedars (*Thuja plicata*). The bark of these trees is too acidic to host the lichen, fungi, and moss that hang out on many Douglas-firs and hemlocks. Notice their neatly tapered trunks and wide, buttressed bases. Their leaves are different from those of the other nearby conifers—look for flat, braided scales that are bright green on the top with white markings resembling bowties or butterflies on the underside. Brush your fingers against the leaves (or gently crush a few in your hand) and you will be treated to their intoxicating smell, somewhere between cedar and spicy pineapple.

There is nothing like time spent in old-growth forests. Look around, learn to identify the trees, and appreciate the carpet of ferns, berries, shrubs, and seedlings far below their towering tops. Breathe in the air. Notice the smells and the sounds. Old-growth forests are a feast for the senses that also speak to something much deeper.

The buttressed roots
of a western red cedar

**Some of the best trails for seeing old growth include Big Beaver Trail, Thunder Creek Trail, Happy Creek Forest Walk (which includes interpretive signage), Horseshoe Bend Trail, Shadow of the Sentinels Trail, and Baker Lake Trail. Always research trails and conditions before setting off.**

# Yosemite

**STATE**
CALIFORNIA

**ANNUAL VISITATION**
4 MILLION

**YEAR ESTABLISHED**
1890

## granite cliffs & famed waterfalls

| | |
|---|---|
| **SUPERLATIVE** | Yosemite Falls (which is actually three falls) is the highest waterfall in North America |
| **CROWD-PLEASING HIKES** | Mirror Lake (easy); Mist Trail (moderate); Yosemite Falls Trail (strenuous) |
| **NOTABLE ANIMALS** | American dipper (*Cinclus mexicanus*); Sierra Nevada red fox (*Vulpes vulpes necator*); western pond turtle (*Actinemys marmorata*) |
| **COMMON PLANTS** | Giant sequoia (*Sequoiadendron giganteum*; see page 56); mountain hemlock (*Tsuga mertensiana*); spider lupine (*Lupinus benthamii*) |
| **ICONIC EXPERIENCE** | Visiting Mariposa Grove to see 1,000-year-old sequoias |
| **IT'S WORTH NOTING** | YOSE is a major climbing destination for climbers from around the world. |

**WHAT IS YOSEMITE?** Yosemite Valley is a glacially carved wonderland tucked high in the Sierra Nevada of California. Snow-fed waterfalls careen down walls of sheer granite up to half a mile high in places. The valley is the centerpiece of YOSE, a huge stretch of mountainous terrain crisscrossed by hundreds of miles of trails but only a few paved roadways. This is one of the canonical places that inspired the establishment of the national parks. Stand between walls of granite, walk among impossibly massive trees, and look up to see some of the roughly two hundred fifty bird species—YOSE will make you feel small in the grandest way possible.

# GEOLOGY

## WHO PUT THAT THERE?

Vandals have been known to push glacial erratics off clifftops, destroying important parts of YOSE's geologic record. As usual, have respect for the landscape and (obviously) don't push boulders off of cliffs.

**GEOLOGY IS ALL AROUND US** wherever we are, but YOSE gives you a real chance to see the results of some major geological processes, no matter which way you look. From the glacially carved Yosemite Valley itself to the towering sheer granite of Half Dome and El Capitan to the countless waterfalls in the park, there's always something extraordinary to see in YOSE. There are smaller and humbler geologic wonders in the park, though. For the past 2 million years, glaciers have advanced and retreated here, leaving plenty of evidence in their wake. Sure, sometimes that evidence is a massive valley. But there are aptly named features, called *glacial erratics*, that have been left behind as well. What looks like a boulder mistakenly plopped down by a crane was actually left behind when a glacier retreated.

During the last 2 million years, which is relatively recent in the Sierra Nevada's geologic history, glaciers advanced and retreated multiple times. There were four notable periods of glaciation here; the most recent is called the Tioga glaciation. It happened between 26,000 to 18,000 years ago and left most of the glacial evidence we see in the park today, including erratics. Think about it—while there are older glacial features in the park, it stands to reason that the Tioga glaciation would have reshaped the land, wiping away much of the evidence of prior glaciations.

The Tioga glaciation left some really lovely erratics throughout the park, especially near Olmsted Point and Tenaya Lake. As a glacier advances, it picks up boulders along its way and/or boulders fall onto it. The glacier brings these along as far as it advances and then when it retreats (melts), the boulders are inevitably left behind. Often, glacial erratics look different from the rock where they've been placed—because they *are* often different types of rock! Color and texture are helpful indicators to the naked eye to determine if something might be an erratic, so be sure to get up close to compare the boulder to the bedrock on which it sits. You may also find erratics that are very similar to the bedrock underneath. This is especially true at Olmsted Point, where some of the erratics are Half Dome granodiorite, just like the bedrock. Simply put, these erratics didn't travel as far as some others.

While erratics from before the Tioga glaciation are less abundant in the park, they're still there. They'll often appear more heavily eroded, since they've experienced much more exposure—some even sit on pedestals of bedrock that eroded underneath them, which makes them look like they could topple over at any moment. However, some of the boulders perched on pedestals in the park aren't erratics, but sit on structures formed in the same way. Whatever their origin, though, these smaller features are more than just boulders, and they're delightful to come across during your time in the park.

A glacial erratic
perched on a pedestal

**The best places to see erratics are Olmsted Point and Tenaya Lake, although they can be found elsewhere in the park. Be sure to ask a ranger for some help finding more erratics!**

# WILDLIFE

## GETTING HIGH WITH BUTTERFLIES

**HIKING THE HIGHER ALTITUDES OF** YOSE is worth the patience and effort it takes to get there—the views from Tuolumne Meadows (an altitude of 8,619 feet) and even higher are extraordinary. The scenic Tioga Road is the only way to get to these elevations by car, and, like most subalpine and alpine roads in the parks, the area is accessible only seasonally. During late spring, summer, and fall, though, visitors are rewarded with ephemeral wildflower blooms and butterfly displays. More than a hundred species of butterfly have been spotted in the park; of which, sixty or so species can be found at elevations higher than 10,000 feet. Even though butterflies don't live very long lives (some for only a few weeks), their presence and changing numbers help scientists understand how climate might be affecting high-altitude habitats.

First, a little lesson on the butterfly life cycle. Butterflies rely on host plants as egg-laying sites. When the time comes, a caterpillar eats its way out of its egg and then begins feasting on its host plant. Once the caterpillar is fully grown, it forms a chrysalis and starts the transition to its adult lifeform. Inside, the caterpillar essentially digests itself (tasty!), leaving behind groups of cells called *imaginal discs*. These cells use the nutrient-rich caterpillar soup to grow wings, legs, and everything else that makes an adult butterfly. Then, lo and behold, a beautiful winged creature emerges to start the process all over again.

Some thirty-five alpine butterfly species in the park are found *only* in the high alpine because that's where their host plants live. It's possible that as the climate continues to warm, their fragile habitat may be reduced or retreat to higher altitudes. Ultimately, it's all a delicate balance. Many plants rely on butterflies to pollinate them; butterflies rely on host plants to sustain eggs and caterpillars; other animals eat alpine plants and butterflies, and so forth. Alpine butterflies can be helpful indicators about climate change, and as studies

## FLUTTERING BY

Here are four examples of beautiful butterflies in a variety of colors that are common in the park.

**Greenish blue (*Plebejus saepiolus*)** Look for green to grayish blue, with a black band near the wing edge. These are the distinctive males; females are brown, and both sexes have a gray to tan underside.

**Clodius parnassian (*Parnassius clodius*)** Look for a translucent yellow-white color, with black spots on the forewing and two orange or red eyespots on each hindwing. They have plump, fuzzy bodies, and males and females look very similar, differing only in some light striping on the females.

**Painted lady (*Vanessa cardui*)** Both males and females are orange with hints of pink and red, with black corners on the forewing. There are usually white spots on each corner as well.

**Orange sulphur (*Colias eurytheme*)** Both males and females are orange and yellow on the upper side. Each wing also has a single eyespot, brown on the forewing and orange on the hindwing. Females have additional dots along the wing edges (pictured).

continue, management and conservation strategies can be implemented to help protect them and their habitats.

Butterflies' short lifespans can mean a very reduced research timeline for scientists to collect annual data. This is where citizen scientists—that's you!—come in. YOSE encourages people to photograph butterflies that they see as they're out hiking, especially in alpine areas. If you do snap a good photo, especially of a rare or unusual butterfly, be sure to note details about the date and location and then visit a visitor center or speak with a ranger to share your wildlife sighting—it could really help scientists with continuing research! There is also an annual butterfly count each summer, starting in Tuolumne Meadow, in which lepidopterists and butterfly enthusiasts unite to count as many species as they see. It's a great day of learning about butterflies in the park, gathering long-term data for research, and building enthusiasm about butterfly conservation. YOSE's higher-altitude regions are well worth the trip; and if you're lucky, you'll see some unforgettable butterflies in action.

Annual butterfly and bird counts happen all over the world.

**If the timing lines up, sign up for the annual Yosemite Butterfly Count. If not, head to Tuolumne Meadows and hike one of the trails in the area.**

## BOTANY

### LOVIN' LICHENS

**ONE OF THE OLDEST PARKS** in the system, YOSE has always been known to inspire enthusiasm for the beauty of nature. It's recognized around the world for its striking granite faces (and amazing climbing), waterfalls, and huge elevation range. It's really easy to get caught up in the vistas, but that can mean that inevitably you miss important flora and fauna. Until 2007 or so, the park had a significant group of residents that went largely unstudied. Once you notice them, though, you won't be able to stop seeing them—lichens! Lichens live at nearly all elevations in YOSE, from the lowest valleys to the highest peaks. And with more than five hundred (and counting!) species in the park, there is a lot to learn from them.

Lichens look to be one entity to the naked eye but are actually two different types of organisms living symbiotically: algae (or sometimes cyanobacteria, called blue-green algae) and fungi. Sometimes the algal partner can live by itself, but the fungus cannot. This partnership allows lichens to grow in all sorts of environments, and they're able to thrive in extremes where most other life doesn't stand a chance. In YOSE, they're found on granite cliffs, tree branches and bark, rocks, and a lot of other seemingly uninhabitable places. They come in a variety of shapes and colors too. One type that you'll likely see is the abundant *Candelaria pacifica*. It's usually yellow, ranging from lemon-yellow to yellow-green, and you'll likely see it on conifers (even dead ones) in the park's lower elevations. Unfortunately, its presence isn't a sign of good things happening in the air.

Lichens live directly off of molecules in the air and moisture around them, so they're very helpful in studying and tracking air quality for a given area.

*Candelaria pacifica* on a branch

Sometimes lichens live with nonlichens, creating colorful displays. You can see this at Bridalveil Fall. The orange color is an iron-based mineral stain, the black colors are lichenized algae and many species of nonlichenized cyanobacteria, and many of the other colors seen on the rocks are crustose lichens.

YOSE's location makes it a particularly great place to examine lichens; since it's protected wilderness, many parts of the park are relatively untouched by human activity. Of course, cars and other vehicles are used in the park, which in turn affect air quality. Perhaps more important, though, nearby areas outside of the park are heavily farmed, utilizing huge amounts of resources, like fossil fuels, that put high levels of nitrogen (among other elements) into the air. This abundance of nitrogen pollution is changing the natural lichen community composition in YOSE. For instance, our yellow friend *Candelaria pacifica* is a huge fan of nitrogen and is thriving in some areas of the park where it hasn't before—a sign that there is too much nitrogen in the air. Presumably, without management efforts, nitrogen levels will either stay the same or rise, which has massive ecological implications. The good news is lichens are providing this helpful information and can be key players in turning that air-quality train around. Maybe as people like you visit YOSE, they can learn about how important conservation is and spread the word.

Lichens in SEKI have been more severely impacted by elevated nitrogen levels than lichens in YOSE.

As you're out looking at lichens, be sure to step lightly and observe respectfully. Get a closer look with a magnifying glass or a loupe—you'll get a detailed view of this magical relationship in action.

**Lichens can be found all over the park and come in a variety of textures and colors. Many lichens can be found together; one place you can see this is Hetch Hetchy when the lake is low. There is a white "bathtub ring" around the lake; but the white is not mineral staining, it is the color of clean rock. The color of the rock above the ring is the combined color of dozens of very common crustose lichen species!**

# joshua tree

**STATE**
CALIFORNIA

**ANNUAL VISITATION**
3.9 MILLION

**YEAR ESTABLISHED**
1994

## enigmatic boulders & iconic plants

| | |
|---|---|
| CROWD-PLEASING HIKES | Barker Dam Nature Trail (easy); Mastodon Peak (moderate); Lost Horse Loop (strenuous) |
| NOTABLE ANIMALS | Red-spotted toad (*Anaxyrus punctatus*); American kestrel (*Falco sparverius*); desert iguana (*Dipsosaurus dorsalis*) |
| COMMON PLANTS | Ocotillo (*Fouquieria splendens*; see page 182); cholla (*Cylindropuntia bigelovii*); Joshua tree (*Yucca brevifolia*) |
| ICONIC EXPERIENCE | Driving Geology Tour Road |
| IT'S WORTH NOTING | Monsoon rains can cause massive flooding, so always check the forecast before you head into the park. |

**WHAT ARE JOSHUA TREES?** The park's namesake trees are actually giant, armored members of the agave family. Their twisted silhouettes define the Mojave Desert on the park's eastern side. In the west, creosote bushes, cholla cacti, and ocotillos mark the hotter, lower-altitude Colorado Desert. JOTR is also known for the spectacular granite formations that punctuate the landscape. Once magma churning underground, these piles of fractured and rounded rocks make for some truly otherworldly scenery. The boulders create world-class climbing opportunities and provide a spectacular backdrop for the park's starry night skies.

california

# BOTANY

## ATTACK OF THE TEDDY BEAR CLONES

**WHOEVER DECIDED TO NAME THE** teddy bear cholla (*Cylindropuntia bigelovii*) had a decidedly dark sense of humor. Not only are these cacti—plants not known for being cuddly—but they are cacti with some of the spiniest spines out there. This is one plant you won't want to examine too closely, but you should definitely take a stroll among them—in closed-toe shoes.

Teddy bear cholla (*Cylindropuntia bigelovii*)

All cacti grow spines (which are actually modified leaves), but the teddy bear cholla grows its spines with extra gusto. Notice that spines cover much of the plant. Their sheer abundance helps shield the plant from harsh desert realities. They help to diffuse and disperse sunlight and heat on the plant, similar to the umbrellas that photographers use to diffuse light. The spines also help keep the drying desert wind away from the plant's flesh. Teddy bears don't need shade or shelter in the desert; they make their own.

A closer look can reveal a papery sheath around the spines—a signature feature of chollas in general. That sheath is a remnant of the outermost layer of cells on a leaf surface, a holdover from long ago. The spines themselves are covered in a series of microscopic barbs—don't look for those. You can't see them with the naked eye, and you definitely don't want to put your eyeballs that close to a spine. Rows of backward-pointing barbs ensure that the spines go in much easier than they come out—sort of like the backward-pointing teeth of *T. rex* and other predatory animals. Chollas also sport tiny, hairlike spines called *glochids*. If you don't want the big spines stuck in you, you definitely don't want these extremely small and very irritating glochids anywhere close to your skin.

So why the long spines, teddy bear? If you visit the park in spring, you may get the chance to see these cacti bloom with pale green or yellow flowers and produce oblong, cup-shaped yellow fruits. Flowers and fruits are often how plants reproduce, but the seeds of this cholla are usually sterile. The plant reproduces by growing a whole new plant from a broken-off piece of an older teddy bear. So by having spines that attach eagerly to anything that brushes against them, chollas can easily hitch a ride to new patches. That ease of transport may have given rise to another of the teddy bear's nicknames: the jumping cholla.

(magnified 350x)

Teddy bear spines are covered in microscopic barbs that expertly cling to whatever brushes against them. Barbs make the spines difficult (and very painful!) to remove.

Pair those spines with an internal structure that lets the plant quickly and safely break off at its joints, and the teddy bear cholla is ready to reproduce like a rock star. Look for smaller, younger teddy bears growing near the base of older, larger plants. Some of those little light-colored stems are new cacti, but they remain genetic clones of their parent plant. These ancient, spiky creatures have a lot to say, so get out there and listen . . . just don't get too close.

It's entirely possible that every teddy bear you see in the Cholla Cactus Garden is actually a clone of the same plant!

**Chollas appear throughout the park but are especially dominant in the Cholla Cactus Garden. Trailside interpretation provides information about the other plants that share the chollas' habitat.**

# GEOLOGY

## FIND YOUR FAULT

**WHEN YOU THINK ABOUT THE** geology in California, it's very likely that faults and earthquakes come to mind. While California isn't the state with the most annual earthquakes, it's still a great place to see evidence of faults and seismic activity.

JOTR's SoCal location makes it an ideal escape from nearby Los Angeles (an easy two- to three-hour drive). Its desert landscape makes it a prime spot to *actually see* evidence and effects of faults—fractures that split between two blocks of rock. Each side of rock moves relative to the other along the fault. The San Andreas Fault is the most studied fault in the world, and for good reason: This fault system where the Pacific Plate and the North American Plate rub up against each other is more than 800 miles long. The earth rumbles as they slide past each other. The San Andreas Fault isn't located within the park boundaries, but you can see it from Keys View.

While the fractures you're looking at may seem unremarkable, consider that this is one of the most active fault zones in the world. These rocks are in motion, moving an average of about 2 inches per year, which means that in a measly 15 million years or so, Los Angeles and San Francisco will probably be neighbors. Hopefully that's enough time for them to set their differences aside.

Hundreds of faults, some of which are branches from the San Andreas Fault zone, run throughout JOTR. The Blue Cut Fault slices right through the center of the park in an east-west direction. It hasn't had any activity in the last 8,000 years or so, but it's one of the longest faults in the park (at least 18 miles) and has left its mark. You can see evidence of the fault's movement in the fault scarps at the base of the Hexie Mountains. A fault scarp usually looks like a stair step (see page 120 for more on fault scarps). If you visit the popular Oasis of Mara, you'll be hanging out around a scarp that is 10 feet high and runs for a half mile.

# FAULTS EVERYWHERE

The JOTR region is littered with faults, some of which are major faults that pass directly through the park. Use this map to understand where you are in the park in relation to the major faults in the area, and think about how some of these faults are likely related to the San Andreas Fault zone.

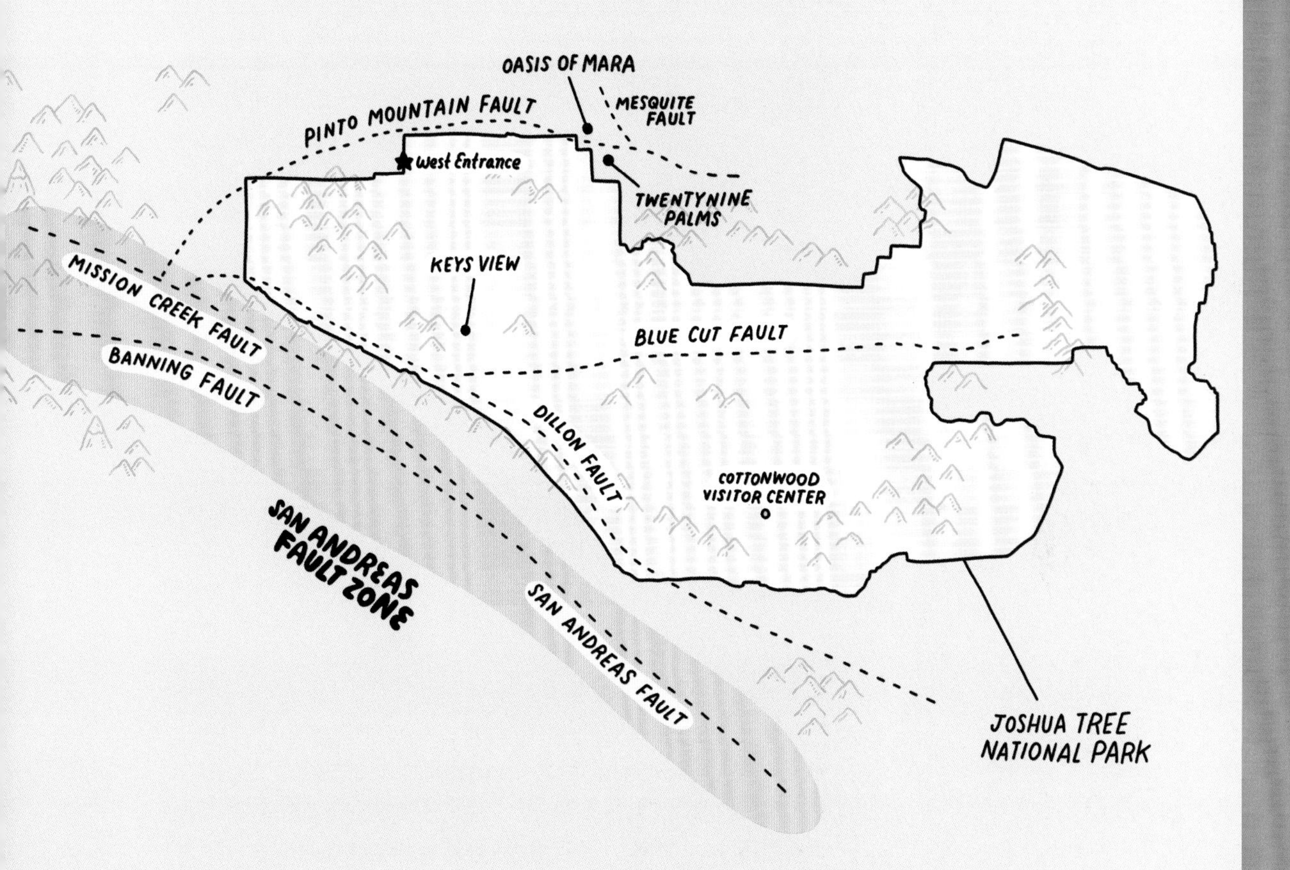

JOTR is known for its oases, which initially might seem out of place in this otherwise arid environment. The movement that happens along faults causes rocks underground to grind into fine powder, creating barriers to water's flow beneath the surface. These underground barriers force groundwater upward, creating an oasis. Five oases in the park are home to the park's largest plant, the desert fan palm (*Washingtonia filifera*), native to Southern California's deserts. They're tall compared to other palms in North America, reaching up to 75 feet. They have huge, accordion-like leaves that provide shade and help make this a cool, safe haven for JOTR's wildlife.

The landscape may seem still, calm, and static, but there is constant activity happening underground. The world under your feet in JOTR is alive with tension and movement.

**You can see the San Andreas Fault from Keys View. You can see oases formed by faults at the Oasis Visitor Center (Oasis of Mara) and 49 Palms Oasis Trail (49 Palms Oasis). Check with a ranger about other oases to see.**

## WILDLIFE

### THE IDEAL DESERT RODENT

**DEPENDING ON WHEN YOU VISIT** JOTR, it might seem like a really inhospitable place for life. Hot days, cool nights, and very little water all come standard in the desert, and it's certainly not an *ideal* place for humans. But like everywhere else on the planet, life has found a way here. Look at the plants dotting the desert—yes, there are even melons (gourds, actually) and flowers among those cacti and shrubs. Animals live among these plants—lizards and insects and tortoises, and one mighty mammal that might be the fuzzy friend best adapted for the unique struggles of the desert.

The kangaroo rat (*Dipodomys* spp.) is a small rodent, with a weight maxing out at 4.5 ounces—no more than a deck of cards—and it spends nearly all of its 2- to 5-year life underground in its burrow. Don't be misled—these rodents are not kangaroos or rats; they are actually most closely related to pocket gophers. They're largely nocturnal, to avoid the heat of the day, but you can spot evidence of them if you look for holes in the ground.

Kangaroo rat (*Dipodomys* spp.)

They typically find homes underneath bushes such as the creosote and various shrubs. Look for mounds of dirt and sand around the base of plant, then look for a series of small holes, golf ball to softball size. Those are the entrances and exits of the k-rat burrow. These burrows protect them from predators and keep them cool during the day, but that's not all they need to thrive in this arid environment.

Here's where things gets crazy. Our dear friends the k-rats do not need to drink water. (We'll wait while you read that again.) They gather seeds in their cheek pouches and store them in their burrows. The seeds absorb moisture from the ground

All mammals metabolize water from their food—kangaroo rats are special because that is the *only* way they get water!

while they're just stashed there, waiting to be eaten. When the k-rats eat these seeds, they get more than food—they're able to metabolize water from the seeds, meaning they don't have to hunt to find a water source.

Their bodies are adapted for this low-water life, and their kidneys are some of the most efficient of any mammal. Instead of peeing super-diluted urine, like we do, they pee a super-concentrated urine that some scientists describe as pasty or crystal-like. They also don't pant or sweat, because that would cause them to lose too much valuable water. Of course, they're always vulnerable as prey to any number of creatures out there, from owls to snakes to coyotes. Fortunately, they have super-hearing, so they can attempt to flee as soon as they hear any sign of a predator. Giant hind legs can send them a whopping 9 feet in one jump, more than twenty-five times their body length. Get out there and look for those holes, and just remember that you *do* need to drink a lot of water in the desert and should NOT be peeing crystals—if you are, get help.

**Look for kangaroo rat burrows in the park, especially near creosote bushes.**

STATE
**CALIFORNIA**

ANNUAL VISITATION
**1.7 MILLION**

YEAR ESTABLISHED
**1994**

# death valley

## quintessential desert & tall mountains

| | |
|---|---|
| SUPERLATIVE | Badwater Basin is the lowest point in North America; Furnace Creek Ranch holds the record for hottest air temperature ever recorded (134°F) |
| CROWD-PLEASING HIKES | Ubehebe Crater (easy); Dante's Ridge (moderate); Mosaic Canyon (strenuous) |
| NOTABLE ANIMALS | Roadrunner (*Geococcyx californianus*); black-tailed jackrabbit (*Lepus californicus*); coyote (*Canis latrans*) |
| COMMON PLANTS | Creosote bush (*Larrea tridentata*); Joshua tree (*Yucca brevifolia*); cottontop barrel cactus (*Echinocactus polycephalus*) |
| ICONIC EXPERIENCE | Exploring the Badwater Salt Flats |
| IT'S WORTH NOTING | The Timbisha Shoshone are the original stewards of this land, and some tribal members continue to live here as they have for at least a thousand years. |

**WHAT IS DEATH VALLEY?** Death Valley is a misnomer. It's not a valley—those are carved by water—but a basin created when the mountains on either side of it rumbled upward. It's also not full of death—not if you know how to comport yourself in the desert. The name is relatively new, coined in the mid-1800s by hapless Euro-Americans who wandered onto the basin floor without adequate water or a meaningful plan. In reality, one of the hottest, driest places on Earth is actually full of life and color. DEVA is a celebration of the endless innovations that plants and animals are capable of, and it's a brilliant reminder that the wonders of the universe are much closer than we sometimes think.

# GEOLOGY

## HELLO FROM MARS

**LET'S BE REAL, YOU PROBABLY** aren't going to Mars. (That isn't to say that humanity isn't going, so please put your energy there if you're called to it!) But you can actually come pretty damn close without ever leaving Earth. DEVA, it turns out, is an amazing analog to the Red Planet.

Of course, the two aren't exactly the same. The sky on Mars is a dusty red and contains no puffy clouds, and there are no plants or soil on the ground. It's much colder there—only in summertime and only at the equator is it ever above freezing. But focus on the bare rocks and sand of DEVA; that's where the Mars parallel is.

As you drive along Badwater Road, keep your eyes peeled for places where water has come pouring from narrow canyons in the mountains. Around the mouths of the canyons are scars on the earth that mark where the water spread out and gradually dropped the rocks and sediments it was carrying—big rocks first, fine sediments last of all. Known as *alluvial fans*, these are a classic sign of water on the move. And there are structures just like them on Mars! Both in DEVA and on the Red Planet, alluvial fans range in size from neatly contained fans only a few feet long to fans many miles in length that are best recognized from above. Thanks to experiments involving the alluvial fans in DEVA, scientists were able to determine that those structures on Mars were probably formed the exact same way. Though DEVA's fabulous alluvial fans weren't the only piece of evidence, they were one key piece in the revelation that Mars, like DEVA, was once a much wetter place.

Wind is another similarity Mars and DEVA share. Gusts full of sand constantly buffet both landscapes—and the rocks are here to tell you about it. Look closely at almost any loose rock in the park and you'll notice it has been polished to a matte sheen. On larger rocks still in place, you might very well be able to see piles of sand on their downwind sides and long, parallel tracks of lines on their surfaces. Notice any rocks with bizarre oblong pits? Those are chunks of basalt, which often cool with gas bubbles trapped inside them. These pits, excavated and enlarged by the wind, can make the rocks look stippled or deeply grooved. All of these features have also been observed on Mars.

## GUSHING FANS

Water rushing out of a narrow canyon creates an alluvial fan on the desert floor as it spreads out and drops its sediments.

But scientists don't come here just to read parallels to the Martian landscape—they also test gear bound for the Red Planet. One place that's been the site of more than one test is the aptly named Mars Hill. Step out of your car on this section of Badwater Road and you might as well be looking out at Mars from the back of a robotic rover such as Spirit. In 2004, Spirit landed inside Bonneville Crater on the Martian surface. Download a picture of the landing site before you get to the park and then compare it to the view in front of you at Mars Hill. The distribution and size of the rocks looks virtually identical to Bonneville Crater! Gaze out at that jumble of rocks and think about the postcard you could write from Mars (TBD what the postage on that would be): *This endless rock field has nice views. Saw a blue sunset. No sign of life yet. XOXO*

**The alluvial fans on the east side of the basin are smaller and easier to see than those on the west, which can stretch for miles. Several especially well-defined alluvial fans can be seen on Badwater Road, just south of Badwater Basin. These features are always easiest to spot from above, so mountain hikes with basin views and some higher-elevation roads are great places to spot them. Mars Hill (no longer marked on park maps) is across Badwater Road from the Artists Drive exit.**

## BOTANY

### MERRILY WE CLONE ALONG

**BECAUSE OF THE UNFORTUNATE NAME** given to Death Valley National Park, you might not anticipate seeing any life at all during your visit. But no matter the season, you'll be able to view some amazing plants thriving here. Even in the scorching heat of summer, you'll see cacti, shrubs, trees, and some species of wildflowers happily hanging out, living that sunny desert life. DEVA gets about 2 inches of precipitation per year, so special sights, like magnificent wildflower blooms, are rare.

All is not lost, though. There is one plant you're sure to see during your time in DEVA, as well as in other parts of the Mojave (and most other North American deserts). It's the Swiss Army knife of desert living and it's called the creosote bush (*Larrea tridentata*). The most drought-tolerant perennial on this continent, it boasts not one but several key adaptations that allow it to thrive in this arid landscape.

Creosotes are humble shrubs that usually grow between knee and head height. You'll learn to easily spot their scrubby profile and tiny, dark green, waxy leaves. If you're around when it rains, you can see them bloom super-cute yellow flowers. If you're wondering whether what you're seeing is a creosote bush, cup some of its branches in your hand, exhale onto them, and then sniff them. You'll smell a distinct, pleasant-but-bitter odor that many from the Southwest United States call "the smell of rain" because humidity activates a protective resin that helps it retain fresh rainwater.

That's not the only clever adaptation of these spectacular shrubs, though. Their root system can grow both vertically (a taproot for reaching water deep in the ground) and horizontally (an extensive system that can quickly absorb rainwater as it soaks into the ground), giving them a huge advantage when rain does fall. Creosote bushes can also go an astonishing *two years* without any rainfall at all.

What might be the most interesting evolutionary strength of the creosote is how it reproduces. These shrubs do have seeds, but it's not surprising that it's incredibly difficult for their seeds to take hold in desert environs. Where does that leave them? They reproduce mostly by cloning themselves, so although individual plants don't seem to live past 150 years (which is still old, folks!), clonal colonies live much longer. The oldest known creosote clonal colony, in the nearby Mojave Desert, is an estimated 11,700 years old (based on carbon-14 dating); it's known as King Clone. You're likely to see colonies that have utilized clonal reproduction and may be thousands of years old themselves. Look for a large mound of sand in the center of a colony, with uniform growth around it. Watch out for cacti because they take advantage of the shade the bushes provide. You might also come across the burrow of a kangaroo rat or a desert tortoise.

Happy clone hunting!

Creosote bush
(*Larrea tridentata*)

**Creosote bushes can be found throughout the park and are especially abundant at lower elevations.**

STATE
CALIFORNIA

ANNUAL VISITATION
1.2 MILLION

YEAR ESTABLISHED
1890

# sequoia

## giant forests & jagged granite

| | |
|---|---|
| SUPERLATIVE | Mount Whitney is the tallest peak in the lower 48 |
| CROWD-PLEASING HIKES | Moro Rock (easy); High Sierra Trail (moderate); Alta Peak Trail (strenuous) |
| NOTABLE ANIMALS | Yellow-bellied marmot (*Marmota flaviventris*); California quail (*Callipepla californica*); black bear (*Ursus americanus*) |
| COMMON PLANTS | Red fir (*Abies magnifica*); meadow lupine (*Lupinus polyphyllus*); limber pine (*Pinus flexilis*) |
| ICONIC EXPERIENCE | Walking among the giant sequoias |
| IT'S WORTH NOTING | A free shuttle runs during the summer through the Giant Forest and Lodgepole areas of the park. Shuttles are also available into the park from nearby towns. |
| SISTER PARK | Kings Canyon (SEKI) |

**WHAT IS SEQUOIA?** Jointly managed with its direct neighbor, Kings Canyon, Sequoia National Park draws its name from the world's most massive trees—giant sequoias. The trees grow only on the western slopes of the Sierra Nevada. It's hard to know which is more dramatic: the towering trees or the rough-cut profile of young mountains. If your neck starts to hurt from all the looking up in wonder, take a trip underground through Crystal Cave—a rare cavern carved through marble, polished smooth by subterranean streams and rivers. Grab your daypack and head out into dramatic High Sierra scenery that you will never forget.

# BOTANY

## SEEING THE FOREST IN FUTURE TENSE

Sequoias' love of water spells trouble in a warming world. Declining snowpack in the mountains here means less water for thirsty seedlings; fewer surviving seedlings means fewer giants.

**WALKING THROUGH A STAND OF** giant sequoia trees (*Sequoiadendron giganteum*) is an unforgettable experience. It's humbling to feel so small and *so young* in the presence of towering, ancient trees. Giant sequoias can live for 3,000 years, but it takes centuries for them to become . . . well, giant. As you explore these wondrous groves, don't just look up, look around; sequoia seedlings and saplings are here too. Take a moment to look for them, and you'll be treated to the secret pleasure of seeing the forest in future tense.

Giant sequoias, despite their seemingly easy dominance, are picky creatures, and their reproduction is far from guaranteed. Millions of years ago, their ancestors lived all over the Northern Hemisphere, but today giant sequoias grow in only about seventy-five groves along a 250-mile stretch on the western slope of the Sierra Nevada. This is a tree that knows what it likes.

Most trees fall into one of two lifestyles: those who love full sun and tolerate drought, or those who need lots of water and so thrive in shadier locales. Sequoias like to have their water and their sunlight too—a sometimes precarious preference. They also have very definite ideas about when and how they'll reproduce. Unlike their closest living relatives, the coast redwoods, giant sequoias will not sprout new growth from the stumps or roots of downed trees. (Only very young sequoias, fewer than 20 years old, will sprout in that way.) Instead, giant sequoias reproduce almost entirely through seeds.

Mature trees release hundreds of thousands of seeds, the size and shape of a flake of oatmeal, usually after the heat from a fire causes the cones to open. Tiny and lightweight, the seeds can end up more than 600 feet—almost the length of two soccer fields—away from the parent trees. The competition for resources is intense, and the odds are not with these seeds; though many might germinate, perhaps only one in a billion will live to become a mature tree.

## LITTLE GIANTS

Giant sequoias can live for thousands of years and grow hundreds of feet tall. But even these giants had to start small.

Huge trees, tiny seeds

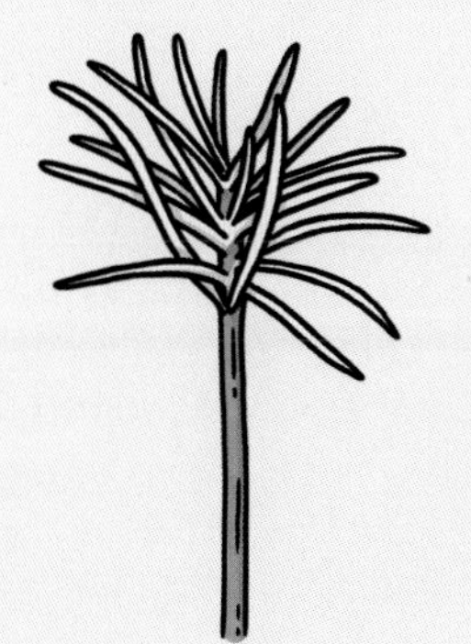

Seedling, newly sprouted and doing its best

Sapling, aka spire top, growing and keeping all of its branches for now

The seeds with the best chance to take hold are those that land in recently burned soil. Fire clears out surface litter and opens up the canopy, allowing light to reach the forest floor and enhance the nutrients in the soil. You will have to look hard to spot the oh-so-tiny sequoia seedlings pushing up from the charred forest floor. Look for tiny plants with a delicate red-purple stalk topped with a few spindly prongs of green several inches long. In some recent burn spots, a carpet of tiny sequoia seedlings could spring up looking quite a bit like grass. Some of the seedlings are eaten by animals; others eventually lose out on the competition for light and water.

The few seedlings that survive then grow into saplings, sometimes called *spire tops* for their pyramidal shape. They will retain their upside-down ice-cream cone shape for the next few hundred years. Look for bark with a purplish tint

Mature sequoias are disease- and fire-resistant. Their sap even contains a chemical, tannic acid, which humans use today in fire extinguishers. Often, lightning is what takes the trees down after a long life of several thousand years.

wrapped around a slender trunk. Unlike mature trees, whose trunks are bare of branches for the first hundred feet or so, the branches of saplings grow along the entire length of trunk, nearly to the ground. Sequoias that haven't yet reached their centennial birthday often sport both living and dead branches.

Sequoias that make it past the spire top stage are considered mature between 500 and 750 years old. Those are the trees you're accustomed to gawking at—as tall as a twenty-five-story building, with bases that could span the average city street. Look for rounded tops and trunks that are bare of branches for the first one hundred feet off the ground.

The chance to be among such awesome giants draws visitors to these groves. Plant your feet, look up as far as you can, and absorb the sounds and smells of the forest. The giants around you all started as thin wisps of plants poking above the forest floor hundreds—or even thousands—of years ago. Their successors are doing the same thing right now; the green shoots of seedlings and the emerging pyramids of saplings have many challenges in front of them. Among those you can see, not many will survive, but those that do are the future of this forest.

**Five groves of giant sequoias are easily accessible in the Sequoia and Kings Canyon parks. Muir Grove is often less crowded than others and well worth the hike.**

## WILDLIFE

### HEAD-BANGING SOCIALISTS

**THERE'S A LOT OF QUIET** grandeur within the borders of SEKI, thanks in large part to the abundance of stately sequoias and jagged granite peaks. While some might call them grand, no one would ever accuse acorn woodpeckers (*Melanerpes formicivorus*) of being part of the parks' quietude. These highly social, extremely talkative birds are the life of the party in the lower elevations of the parks.

You'll likely hear their raucous jabbering before you see any acorn woodpeckers. Listen closely and you'll soon begin to recognize their most common call: a loud waka-waka-waka squawk that sounds eerily like a rhythmically squeaking bed frame. It's how they greet members of their family, argue with the neighbors about territorial boundaries, and generally take care of any interspecies quibbling. You might also hear a loud, single-note squawk or a call that sounds like "carrot-cut" with a rolled *R*.

Acorn woodpeckers store their namesake food in custom-pecked granary trees.

Given their sheer numbers, you will catch sight of an acorn woodpecker sooner or later. They are mostly black, with a red cap, cream-colored face, and a straight black bill surrounded by black feathers. Some people think their markings resemble clown makeup. Acorn woodpeckers live in incredibly complex and large, multigenerational family groups. There might be several breeding males and females in each group, and all work together to raise the young and gather, store, and defend food for the family. Young woodpeckers stay with their families and help to raise the next generation.

Acorn woodpeckers take their name from the food source that gets them through the winter and occupies much of the family time: acorns. Each member of the family chips in to gather them. Then they use their woodpecking skills to

Seeds are gathered and stored year-round, but if you visit SEKI in the warmer months you may also be treated to the sight of acorn woodpeckers dipping and swinging through the air as they catch their favorite seasonal treats: flying insects.

High-speed camera studies have shown that woodpeckers consistently vary the path of their pecks, ensuring that no one part of the brain or skull receives more than its fair share of impacts.

store the food in a granary tree. You'll know instantly when you see a granary tree—thousands of holes give them away. The birds tuck away their summer bounty of acorns in the holes and then eat the nuts throughout the winter when other food is scarce.

Like all woodpeckers, acorn woodpeckers have skulls loaded with adaptations that help distribute and dissipate the force of repeated impacts: sections of spongy, plate-like bone; a brain taller than it is wide; slightly different-sized upper and lower bills; and a modified bone that wraps all the way around the skull like a seatbelt. It's known as a hyoid bone, and you have one too—but yours is a small horseshoe-shaped bone in your neck that helps you move your tongue, not headbang.

Think of acorn woodpeckers as nature's little headbanging socialists. Of course, they don't actually have a system of government or any known inclination toward heavy-metal music (dear scientists, please conduct this study ASAP), but they are all about that communal lifestyle. The young? Raised together. The food? Enough gathered and stored for everybody. The wood? Pecked with safe abandon. If you lived this lifestyle, wouldn't you constantly be squawking about it too?

**Acorn woodpeckers are exceptionally common year-round in the Foothills area just inside the entrance to Sequoia National Park. You'll see or hear them in any forest that includes some oak trees.**

STATE
CALIFORNIA

ANNUAL VISITATION
1.2 MILLION

YEAR ESTABLISHED
1940

# Kings Canyon

## plunging canyons & colossal trees

| | |
|---|---|
| SUPERLATIVE | The Redwood Mountain Grove is the largest remaining natural grove of giant sequoias in the world |
| CROWD-PLEASING HIKES | Big Stump Trail (easy); Big Baldy Trail (moderate); Don Cecil Trail (strenuous) |
| NOTABLE ANIMALS | Pika (*Ochotona princeps*); western tanager (*Piranga ludoviciana*); gopher snake (*Pituophis catenifer*, see page 150) |
| COMMON PLANTS | Blue oak (*Quercus douglasii*); giant sequoia (*Sequoiadendron giganteum*; see page 56); oval-leaved buckwheat (*Eriogonum ovalifolium*) |
| ICONIC EXPERIENCE | Taking in sweeping views of the Sierra Nevada |
| IT'S WORTH NOTING | The Generals Highway connects Sequoia's Foothills Visitor Center with Kings Canyon's Grant Grove. Overlooks along the way provide stunning vistas. |
| SISTER PARK | Sequoia (SEKI) |

STAMP PASSPORT HERE

**WHAT IS KINGS CANYON?** Kings Canyon is a wide glacial valley tucked in between towering peaks. It's one of the deepest canyons in the United States, and it's framed by some of the most dramatic mountain peaks in the country: the Sierra Nevada. This is a landscape attuned to size and drama. Kings Canyon National Park is right next door to and jointly managed with Sequoia National Park. From grueling elevation gains to stunning roadside overlooks to wildflower-strewn meadow strolls, Kings Canyon is an alpine adventure that's not to be missed.

# GEOLOGY

## STANDING UNDER ICE

**SEVERAL TIMES OVER THE PAST** 10 million years, massive sheets of ice have formed in the high reaches of the Sierra Nevada and descended into the land below. Ice thousands of feet thick sculpted and modified the landscape, carving jagged mountain peaks and steep-sided canyons, leaving behind rubbly piles of debris called *moraines*. The glaciers also left behind thousands upon thousands of lakes. They are wonderful hiking destinations and photo opportunities, but they are also a map of the thickest and most powerful ancient ice in these mountains.

Glaciers form in places where snow accumulation outpaces snow melting. Snow compacts into ice and, with enough weight accumulated, the glacier begins to flow downhill as an erosive river of ice. As the glacier grows and descends lower in elevation, it eventually reaches a point at which the snow and ice melt or otherwise disappear faster than new snow can be added. Scientists call the upper zone of the glacier the *snow and ice accumulation zone*. The lower portion, where most melting occurs, is called the *ablation zone*. The lakes scattered all through the higher elevations of the Sierra Nevada roughly mark the line between accumulation and ablation during the last glaciation.

Glaciers are not indiscriminate bulldozers that flatten everything in their paths; think of them more as indelicate landscapers. In parts of the mountains where the ice was thickest, glacial weight and movement fractured the bedrock, exposing harder igneous rocks below. When the ice began to melt back, it revealed numerous depressions in the fractured bedrock. These depressions, bordered by the more resistant rock, captured remnants of glacial ice in addition to newly fallen snow and rain, becoming the lakes you see in SEKI today.

If lakes mark the thickest portion of recent glaciers, giant sequoia groves—which require deep, rich soils to grow—mark places the ice never visited.

Stand next to a lake in the park and look up. If you happen to be visiting on a sunny day, all the better. Let your eyes unfocus and, for a moment, see if you can't resolve the blue

## MAPPING ICE WITH WATER

In the parks, the current locations of lakes (black dots) mirror the boundaries of ice (orange shading) from at least one of the most recent glaciations in the region, known as the Tahoe Glaciation. Dates are still hotly debated, but the end of this period is generally set somewhere between 70,000 and 50,000 years ago.

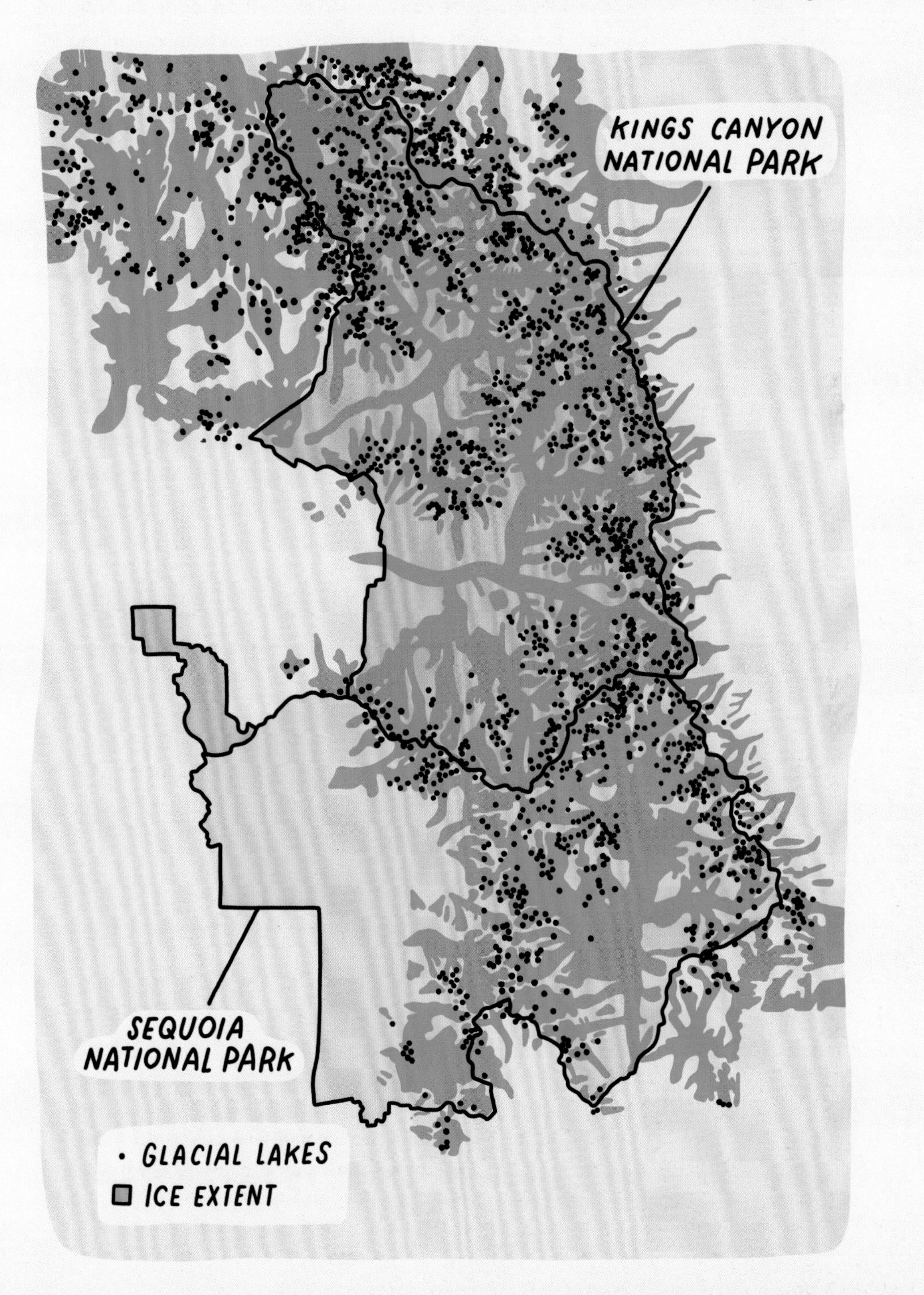

sky into the electric-aqua-blue of glacial ice. Any time you are near a mountain lake in SEKI, you are standing in a spot that would have been covered in thousands of feet of ice in the very recent geological past. Look at the mountains surrounding you; only the very tops and highest ridges would have been able to peek (pun intended) above the ice line. Look around and see if you can spot other glacial calling cards, like striations carved into the rock. Below you, Kings Canyon and the other river valleys would have been filled with far less ice, but still enough to smooth and widen their sides into the U-shaped valleys that exist today.

A *nunatak* is a mountain peak or other high feature that remains exposed above a glacier. The truly intrepid can see some fine examples of them in the Harding Icefield in KEFJ and the Bagley Icefield in WRST.

High mountain lakes are undeniably beautiful, especially in the rugged wilderness of SEKI. There are thousands of them tucked beside hiking trails and below mountain crests. When you visit one, you are visiting the past home of ancient ice powerful enough to rewrite landscapes. The string of lakes that bedazzles the high country here is the X-marks-the-spot on a treasure map of past landscapes and powerful forces.

**Some day hikes and many more overnight backpacking adventures will get you up into the lake-studded wilderness of SEKI. Check in at a visitor center for routes that match your time, preparation, and skill.**

STATE
CALIFORNIA

ANNUAL VISITATION
499,000

YEAR ESTABLISHED
1916

# lassen volcanic

## bubbling mudpots & alpine meadows

| | |
|---|---|
| SUPERLATIVE | Mount Lassen is the site of California's most recent volcanic eruption |
| CROWD-PLEASING HIKES | Bumpass Hell (easy); Devils Kitchen (moderate); Ridge Lakes (strenuous) |
| NOTABLE ANIMALS | Golden-mantled ground squirrel (*Callospermophilus lateralis*); California tortoiseshell butterfly (*Nymphalis californica*); mule deer (*Odocoileus hemionus*) |
| COMMON PLANTS | Jeffrey pine (*Pinus jeffreyi*); lodgepole pine (*Pinus contorta*); California corn lily (*Veratrum californicum*) |
| ICONIC EXPERIENCE | Driving the park highway is an excellent introduction to the scenery |
| IT'S WORTH NOTING | Snow arrives early and stays late in this mountainous park. For the warm-blooded, skiing and snowshoeing offer a chance to explore California at its snowiest. |

**WHAT IS LASSEN VOLCANIC?** Mount Lassen, also called Lassen Peak, is the southernmost volcano in the Cascade Range. It's just one of many volcanoes within LAVO—in fact, it's possible to explore all four types of the world's volcanoes (shield, composite/stratovolcano, cinder cone, and plug dome) within the park. Past eruptions left behind colorful soils and shiny black lava beds. Future eruptions will come too, though it's impossible to say when. In the meantime, roiling heat below ground powers hydrothermal features all over the park. Come for the volcanoes, but stay for the high-country scenery that boasts some of the darkest night skies in California.

# GEOLOGY

## THE SWEET SMELL OF HELL

Recent evidence suggests that high-atmosphere clouds on Uranus are composed mostly of rotten-smelling $H_2S$. The jokes practically write themselves.

**THERE'S A LOT TO DELIGHT** the senses at LAVO: sweeping mountain vistas, carpets of spring wildflowers, and the cool clear air of high altitude. One thing that may not delight you, but will certainly be a part of your sensory experience in the park, is the unmistakable smell of sulfur.

In LAVO's thermal areas, the unpleasant tang of sulfur pervades the air. If you've visited other volcanic national park sites, like the geyser basins in YELL or Kīlauea in HAVO, you're familiar with the smell. In fact, it accompanies many types of volcanic activity. The simple explanation for the close link between sulfur and volcanoes is that there is quite a lot of sulfur in the mantle of our planet. Sulfur melts and mixes with magma underground, becoming one of a number of gases trapped inside the liquid rock. As the magma rises, decreasing pressure releases those gases, including sulfur. In LAVO, sulfuric gases pour out of the ground at gurgling mudpots, boiling pools, and smoking fumaroles—vents that disgorge volcanic gases and steam.

You can see evidence of sulfur in the bright-yellow ground around many hydrothermal features in the park. That is elemental sulfur that has condensed out of the water vapor venting from underground. Really, though, your sense of sight takes a backseat when it comes to sulfur. This is one element that the nose knows. Have a short but dedicated sniff near the thermal areas and you will almost instantly recognize the hydrogen sulfide ($H_2S$) smell of rotten eggs. Summon the courage for a second sniff and you may detect a sharp, pinching smell reminiscent of a just-struck matchstick—sulfur dioxide ($SO_2$). Pay close attention to the plumes of water vapor rising from the ground—when the sun hits just right, you may be able to discern a brown or bluish tint: evidence that the plume is loaded with $SO_2$. In contrast, $H_2S$ is completely colorless.

These gases deposit the yellow streaks of sulfur you can see on the ground, but they don't stop there. Look around the park and you will quickly notice the preponderance of light-colored rocks streaked with red and brown. Those rocks started out as dark-colored lava. Sulfur gases bubbling up from underground combined with water on the surface and produced sulfuric acid that weakened and altered the hard, dark rock into light-colored clay stained with iron minerals.

Sulfur seems noxious to us—it's the "brimstone" of fire and brimstone after all—but we are weak and uninteresting compared to the microbes that thrive in LAVO's superheated, sulfur-laden waters. Many of these bacteria and archaea (possibly the oldest form of life on Earth) are actually using the toxic-to-us compounds dissolved in the water as their source of energy. Mixed in are many kinds of algae that manufacture their energy the old-fashioned way via photosynthesis. These microbes, and the fossilized remains they leave behind, are opening up all sorts of wild possibilities for finding evidence of life beyond Earth. (See page 110 for more on microbes' connection to the hunt for extraterrestrial life.)

You will probably smell the thermal areas in LAVO before you see them. Wrinkle your nose all you like; the fact remains that the sulfur you can see and smell here connects us to the depths of the earth and the farthest-flung planets of our universe.

The human nose can actually detect $H_2S$ in much smaller quantities than any scientific instrument, although we quickly become accustomed to it and thus fail to recognize when the gas has reached dangerous levels.

**Of the eight hydrothermal areas in LAVO, Sulphur Works and Bumpass Hell are the most easily accessible. Devils Kitchen, reached via a short hike, offers mudpots, fumaroles, and hot springs with fewer people. Hydrothermally altered rocks occur throughout the park. Always use caution and stay on the boardwalks where provided.**

# WILDLIFE

## BIRD-BRAINED FORESTRY

Clark's nutcrackers have a special relationship with whitebark pine trees.

**FLASHES OF GRAY AND BLACK** populate forests, campsites, and roadside pullouts in LAVO. Clark's nutcrackers (*Nucifraga columbiana*) are a common sight throughout western North America, but these medium-size members of the crow family might as well be issued ranger hats and forestry PhDs because of the major role they play in shaping the pine-dominated forests of LAVO.

As far as avian life goes, Clark's nutcrackers are on the decidedly less flashy end of the spectrum. They've gone for a look you might call understated but carefully considered. Mostly gray, they have black wings and tails with a white underside that accentuates and perfectly coordinates with their long, finely tipped black bills.

Take a moment to consider that bill. It's the key tool that makes the Clark's nutcracker invaluable to the forests here and throughout western North America. The birds use that lance-shaped bill to crack open pine cones and extract the nutritious seeds. Though nutcrackers will eat just about anything—including carrion and the occasional smaller bird—pine seeds are the key component of their diets. In fact, the birds spend so much of their time wrenching cones open that you can sometimes see reddish purple patches on a bird's gray chest feathers, dyed by the cones' sticky resin.

The ability to pry open still-sealed cones gives the birds access to a large amount of nutrients that other forest dwellers miss out on. Even after a bird has eaten its fill, it will continue to collect seeds and store up to one hundred fifty of them at a time in a specialized pouch under the tongue. Those seeds, the bird's winter food source, wind up distributed throughout the forest a few at a time, often as far as 20 miles away from the nest site. A single bird caches tens of thousands of seeds over the course of the season, and during peak times may collect as many as five hundred seeds an hour.

If you're lucky enough to spot a nutcracker caching its seeds, you may see the bird dig a shallow trench in soil or push individual seeds into decaying wood or loose, gravelly soil. They can remember thousands of individual locations with stunning accuracy. Sometimes, though, caches get left behind, and that's when the nutcracker's forestry work takes hold. They bury the seeds at just about the ideal depth for germination, which means that, given time and luck, abandoned seeds will sprout into new trees. By caching seeds across a huge territory, the birds help spread trees between isolated patches of forest—incredibly important in an age of increasingly fragmented habitats and changing climate.

For one species of tree, the work of these birds is absolutely critical. The cones of the whitebark pine tree (*Pinus albicaulis*) do not open by themselves. The trees, which have been hit hard by an invasive fungus called blister rust, as well as the already apparent effects of global warming, depend almost entirely on Clark's nutcrackers to distribute their seeds. The birds eat many types of seeds, but none are as calorie packed as those of the whitebark pine.

Cached seeds mean that Clark's nutcrackers don't have to migrate to warmer climes during winter; you can see them in LAVO year-round. With the exception of caching, which they do alone, you can usually find them in groups, sometimes with other birds and mammals hanging out nearby as well. Nutcrackers are like the cool kids who have snacks no one else in the forest does; who wouldn't want to hang out with them?!

Clark's nutcrackers have understated coloring and lead relatively quiet lives that may not always attract much attention. But take some time to observe a few; they are among the most influential architects in this forest.

In developed areas, these birds often beg humans for food. You must resist! Feeding wildlife harms the animals and puts you at risk for disease or injury.

**Clark's nutcrackers are common throughout the park in all seasons.**

## MOUNT TEHAMA: MISSING MOUNTAIN

**THERE'S SO MUCH** to see along the park highway that it might surprise you to know that one of the most interesting things along the road is something you *can't* see. At the park's southern border, the highway cuts directly through the middle of what was once an 11,000-foot-tall stratovolcano, Mount Tehama. The volcano started to rise around 600,000 years ago, went silent after 200,000 years of eruptions, and was aggressively eroded by precipitation, glaciers, and hydrothermal activity. Today, Brokeoff Mountain, Pilot Pinnacle, Mount Diller, and Mount Conard are the most obvious remains of Tehama.

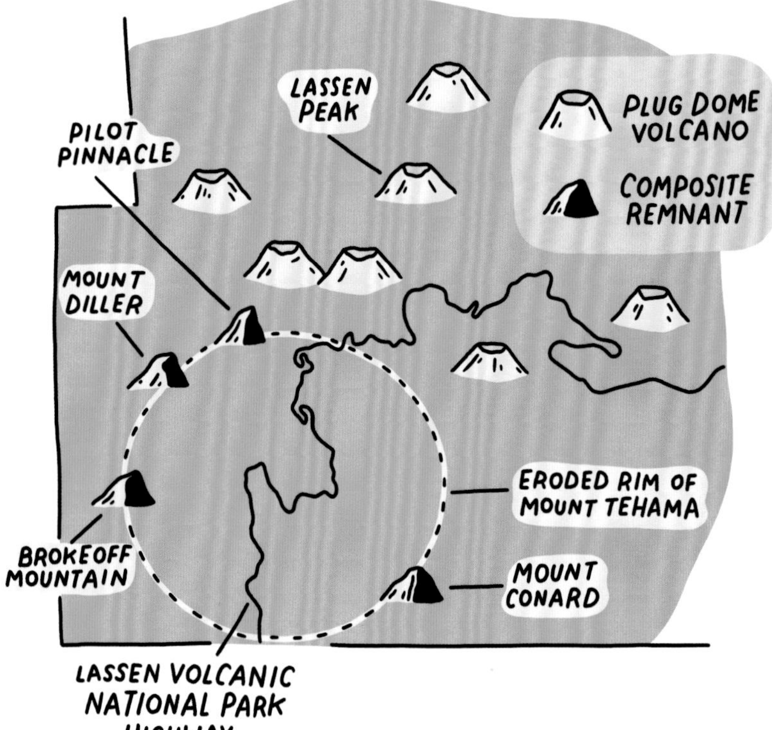

STATE
CALIFORNIA

ANNUAL VISITATION
483,000

YEAR ESTABLISHED
1968

# redwood

## coastal giants overlooking the pacific

| | |
|---|---|
| SUPERLATIVE | REDW is home to Hyperion, the tallest tree in the world |
| CROWD-PLEASING HIKES | Circle Trail (easy); Tall Trees Grove (moderate); Last Chance Section of the Coastal Trail (strenuous) |
| NOTABLE ANIMALS | Great horned owl (*Bubo virginianus*); gray whale (*Eschrichtius robustus*); black-tailed deer (*Odocoileus hemionus*) |
| COMMON PLANTS | Coast redwood (*Sequoia sempervirens*); Sitka spruce (*Picea sitchensis*); sword fern (*Polystichum munitum*) |
| ICONIC EXPERIENCE | Taking a hike to Tall Trees Grove to be among the redwoods |
| IT'S WORTH NOTING | Redwood National and State Parks are coadministered by the NPS and California State Parks. |

**WHAT IS REDWOOD?** Coast redwoods (*Sequoia sempervirens*), the tallest trees in the world, grow only along a narrow strip of the US Pacific Coast. More than 95 percent of them were logged during colonization, but REDW protects about half of all the redwoods left in the world. Even if you haven't been here, you've probably seen it—it was featured in *Return of the Jedi* and *The Lost World: Jurassic Park* due to its lush, untouched greenery. Because it's a coastal park, you'll be able to get near the water to wander the tide pools and, if you're lucky, you might see some harbor seals and sea lions on the rough, rocky shores. Whether you're a landlubber or ocean aficionado, REDW has something special in store for you.

california

# BOTANY

## REDWOODS AND CARBON—A LOVE STORY

**HIKING THROUGH A FOREST OF** coast redwoods (*Sequoia sempervirens*) is a humbling and iconic experience in REDW. The trees grow hundreds of feet high and can easily live 500 years. In fact, the tallest tree on the planet, Hyperion, is a coast redwood in REDW that is nearly 380 feet tall—slightly longer than an American football field. Given the right conditions, some coast redwoods can live for thousands of years. While it may be hard to imagine, there used to be exponentially more redwoods in this region, prior to excessive logging and development. It's estimated that more than 95 percent of old-growth redwoods were lost, and now important work is being done to help redwoods reclaim the land.

Redwoods are synonymous with California. Thanks to the ideal coastal climate, they're better able to thrive here than anywhere else in the world. These giants are thirsty, and they do best when there is regular seasonal rainfall. They're also able to absorb fog from the air, which helps them survive dry spells. Water isn't all that these trees are drinking, though. All trees absorb carbon dioxide ($CO_2$) during photosynthesis, and they store some of that $CO_2$ in their leaves, roots, and trunks. This magical storing process is called *carbon sequestration*, and it's one of the things that redwoods do better than pretty much every other plant on the planet. As human beings continue to pump massive amounts of $CO_2$ into the atmosphere, warming the planet, that carbon sequestration may be a valuable tool in fighting climate change. Because redwoods grow quickly and continue adding significant growth even after they're centuries old, they're able to store more and more carbon dioxide. It's worth noting that redwood root systems are shallow yet expansive, extending laterally from the tree's base over a 50-foot circumference. An entire forest of ancient redwoods can store at least three times more carbon above ground than any other forest on Earth, which makes managing old-growth forests incredibly important. It also means that there is a hefty incentive to properly manage young redwood forests now so that they become the old-growth forests of the future.

Redwoods are very resistant to rot, which means that they retain carbon long after they die.

Protecting these forests for the future is a priority for Redwood National and State Parks, and they're partnering with Save the Redwoods League to take major steps in restoring and maintaining them over the next several decades. When land was logged starting in the mid-1800s the redwoods were replaced with dense populations of other trees, like Douglas-fir, altering the forest's ecosystem. About 40,000 acres of old-growth redwoods are protected at REDW. The mature growth will be thinned and maintained, and second-growth forests of Douglas-fir will be thinned to allow for more biodiversity. The goal is to connect the old growth with the younger forests and make the entire area a diverse forest. It's messy, complicated work that also involves challenges such as relocating streams and removing unused logging roads, but ultimately the restoration has major implications for fighting climate change. As you're hiking among the ancient giants, imagine future generations enjoying an even more expansive and productive forest than you see here today.

**There are plenty of short walks to and through old-growth forests in the park, including Stout Grove Trail and Lady Bird Johnson Grove Trail. Ask park staff for more ways to see these amazing trees, and find out where current revegetation is happening.**

# WILDLIFE

## TAGGING ALONG WITH GRAY WHALES

**SEEING WILDLIFE IS ONE OF** the most-desired experiences on any visit to a national park. However, wildlife is notoriously difficult to plan around. Many creatures are elusive and prefer to hang out in the less-trafficked areas in the parks (who wants to hang out with humans, anyway?). Fortunately for visitors to REDW, there are some massive mammals that can reliably be seen twice a year during their migrations. Gray whales (*Eschrichtius robustus*) make one of the longest round-trip migrations (about 12,500 miles, the equivalent of hiking all the trails in the park sixty-two times) of any animal on the planet.

Approximately 20,000 gray whales pass by REDW twice a year. From March to June, they make the trip from the Baja California Peninsula to the Arctic to enjoy ample food. From September to January, they make the reverse trip to breed and rear their young. Since they travel en masse during these months, you're likely to see a gray from the shore if you know what you're looking for. First of all, these massive mammals can get up to a whopping 45 feet long, the equivalent of three car lengths. Don't look for a dorsal fin because gray whales don't have them. Look for their large, dark bodies at the surface of the water, where they're likely to spout, or blow, warm, moist air from their blowholes. You might notice that they have blotchy skin from the barnacles (see page 13) that make their permanent home on the whales' bodies. Females travel with their young calves and tend to be closer to the shore than their male counterparts during their spring trip to the Arctic. You might also see just the whale's head bobbing out of the water—it's called *spy-hopping*, and apparently it helps the whales get a better or different view of what's above the surface.

Something that you won't see is the whale's underwater life. Grays belong to a family called baleen whales because they feed with baleen: a fringy plate attached to the upper jaw, in lieu of teeth, that filters out water and traps prey such as

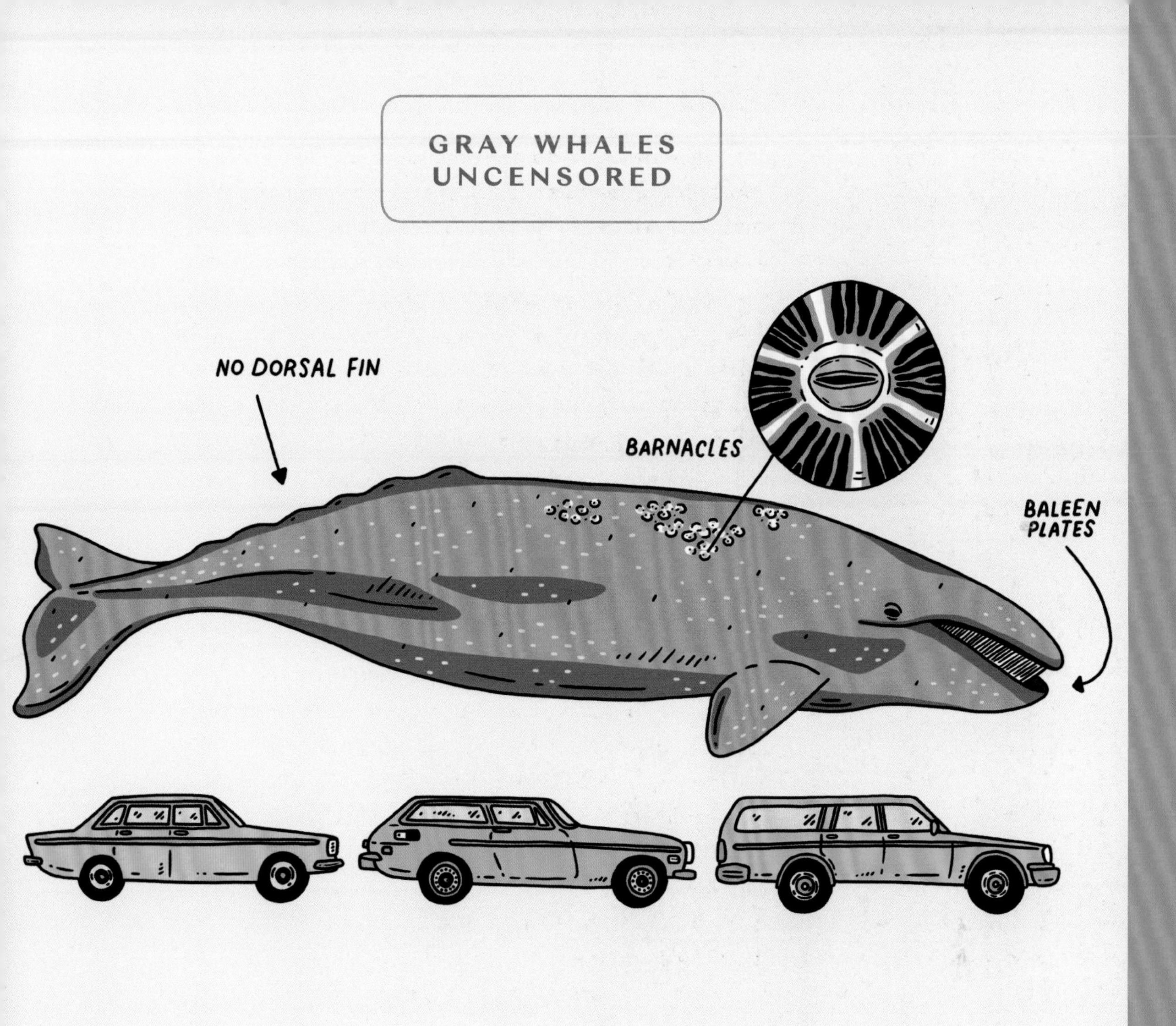


shrimp and other small invertebrates. They eat a *lot* of those critters, though—usually a few hundred pounds per day. The Arctic seas have abundant food for whales on the ocean floor, which is great, because they prefer to feed in shallow waters rather than the open ocean. Interestingly, nearly all gray whales feed on their right side, and they're the only type of whale with asymmetrical baleen plates.

Some grays don't make it all the way to the Arctic with the rest of the group—a couple hundred whales, known as the Pacific Coast Feeding Aggregation, stop between northern California and south-east Alaska, for either a short period or the entire feeding season.

As you're looking at these behemoths, you might forget that they're mammals, just like us. In fact, all whales share a common ancestor, and that ancestor actually lived and walked on land. Over time, that creature spent more and more time in the water and made adaptations for under-water living over millions of years. Modern whales still have remnants of that land-dwelling past, including a pelvis that isn't connected to legs (you know, because they don't have legs). Give your hips a shake in solidarity with these whales as they make their journey, and send them on their way with gratitude for giving you a little glimpse of their watery pilgrimage.

**During gray whale migrations, Crescent Beach Overlook, Wilson Creek, High Bluff Overlook, Gold Bluffs Beach, and Thomas H. Kuchel Visitor Center are all good places to perch with binoculars and scan the waters. You also may be able to see grays any time of the year at Klamath River Overlook. Early morning is typically the best time to see them.**

STATE
**CALIFORNIA**

ANNUAL VISITATION
**366,000**

YEAR ESTABLISHED
**1980**

# channel islands

## a wilderness past the edge of the continent

| | |
|---|---|
| CROWD-PLEASING HIKES | Anacapa Island Loop Trail (easy); Cavern Point Loop (moderate); Montañon Ridge Loop (strenuous) |
| NOTABLE ANIMALS | California sea lion (*Zalophus californianus*); California brown pelican (*Pelecanus occidentalis*); island fox (*Urocyon littoralis*) |
| COMMON PLANTS | Torrey pines (*Pinus torreyana*); giant coreopsis (*Coreopsis gigantea*); island morning glory (*Calystegia macrostegia*) |
| ICONIC EXPERIENCE | Hearing the chatter of seals and sea lions on the way to Anacapa Island |
| IT'S WORTH NOTING | Sea caves riddle the shoreline of several islands and are popular destinations for confident kayakers. |

**WHAT ARE THE CHANNEL ISLANDS?** The Channel Islands are an archipelago off the coast of Southern California; CHIS protects five of the islands and some of their surrounding seawater. Other islands and nearby marine areas are owned and protected to various degrees by a patchwork of organizations. Reachable only by plane or boat and surrounded by sometimes turbulent seas, CHIS is a world unto itself. There's so much to explore on foot or by water; just be prepared to fend for yourself. Island isolation has produced a host of wonderfully strange plants and animals that live high atop rocky cliffs or in the submerged wonderland offshore and everywhere in between.

# BOTANY

## A RAIN FOREST BENEATH THE WAVES

Since it's a type of algae, kelp isn't a true plant.

For a closer, slightly drier look at kelp, keep an eye out for bits washed ashore by rough water.

**THE ROCKY, WINDSWEPT CHANNEL ISLANDS** don't support the same type of towering forests as many places in mainland California. But there are forests filled with giants here—you just have to know where to look. Gaze out at the water from atop the islands, or peek overboard from your kayak, and you'll see them: forests of kelp rising from the ocean floor.

All along the western coast of North America, towering stands of kelp, a kind of algae, sway in the ocean currents. In CHIS, as in most of Southern California, giant kelp (*Macrocystis pyrifera*) dominates these forests. It's one of the fastest-growing "plants" on Earth, adding 4 to 10 inches a day. In ideal circumstances that number leaps to more than 2 feet a day. Kelp easily grows more than a hundred feet tall, about twice as tall as the letters on the famous HOLLYWOOD sign.

Specific parts of kelp resemble familiar structures on terrestrial plants, but with fun and unique differences. Anchoring the kelp in place are structures called *holdfasts*. Though they look like plant roots, holdfasts don't transport water or nutrients—their sole purpose is to anchor the plants in place. *Stipes*, which look like stems, also don't transport nutrients. They form a tough but flexible support system that allows kelp to bend in response to water's movement while moving ever closer to the light-filled surface. Leaf-like *blades* are the engines of kelp; they capture sunlight for photosynthesis and absorb nutrients from the rich seawater stew. Around them, *gas-filled bladders* keep the plant afloat and reaching for the sunlight.

The kelp forests around CHIS have been continually cataloged since the 1980s—the longest-running monitoring program in the National Park Service. More than a thousand species of plants and animals inhabit the kelp forests here, where northern and southern species mingle in warm currents from the south and cooler water from the north.

Much like tropical rain forests, kelp forests host layers of life. The canopy near the water's surface is dominated by small fish and invertebrates, many of which eat the kelp or use it to hide from predators. Look for seabirds waiting for a chance to snatch some of the larger fish that hang out in the mid-story to capture canopy creatures. The forest floor is home to a brightly colored menagerie that includes sponges and urchins. Like seabirds, seals and sea lions are temporary residents that come looking for a meal or to escape rough water or predators.

As in any ecosystem, interfering with one resident can have dramatic effects. Sea otters (*Enhydra lutris*), important and iconic kelp-forest residents, were hunted to the brink of extinction in the early twentieth century. Though they've returned to many California kelp forests and even to some of the Channel Islands outside of the national park, they have not yet returned to CHIS. In their absence, an explosion of sea urchins (a favorite otter snack) overgrazed and killed huge patches of kelp. Today, recovery of other urchin eaters, like spiny lobsters and the California

## FOREST GIANTS

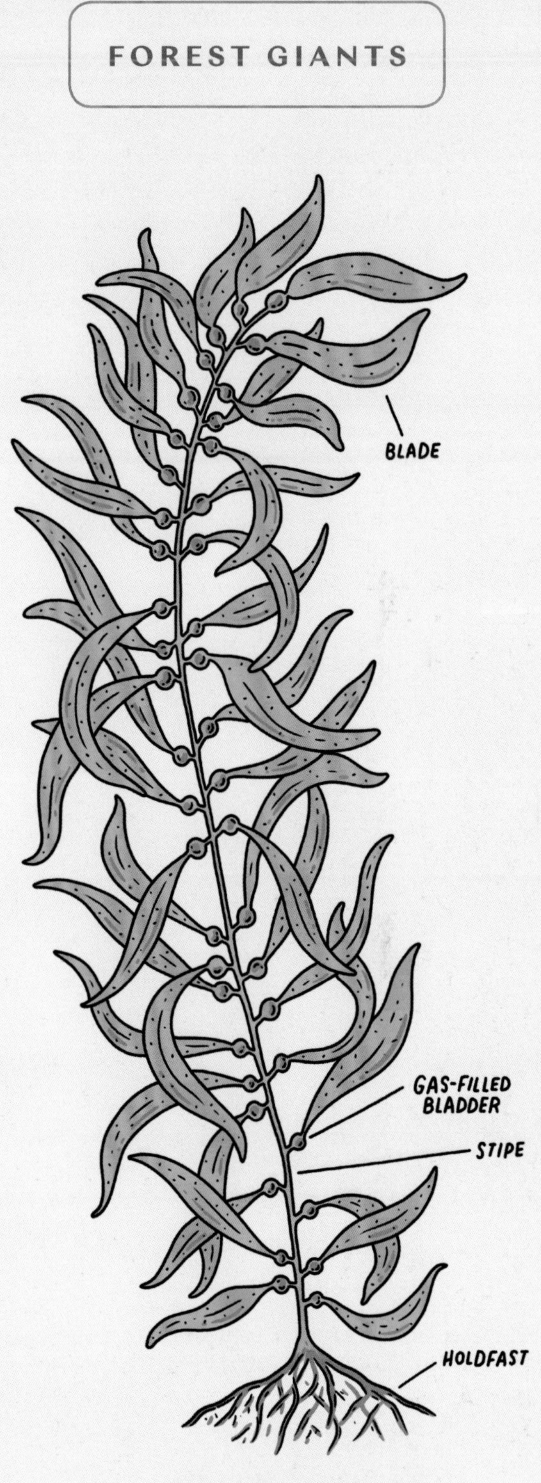


Giant kelp (*Macrocystis pyrifera*)

Sunflower starfish were another important urchin predator. Their numbers plummeted during a disease outbreak in 2014 and have yet to recover.

sheephead fish (*Semicossyphus pulcher*), has helped to slow the spread of urchin barrens.

The kelp fronds you see swaying in the waves around CHIS (and in the patchwork of other protected areas around the islands) are the canopy of a unique world, the tippy top of an underwater forest every bit as intricate and impressive as California's terrestrial tree kingdoms.

Kelp forests can be seen from many points on the islands as well as from boats on the water. Those who want a total immersion can snorkel or scuba in several of the forests. During the summer months, rangers make regular dives in the forest with cameras that broadcast online and in the park.

STATE
CALIFORNIA

ANNUAL VISITATION
222,000

YEAR ESTABLISHED
2013

# pinnacles

## talus caves & volcanic formations

| | |
|---|---|
| CROWD-PLEASING HIKES | Prewett Point Trail (easy); Moses Spring to Rim Trail Loop (moderate); Chalone Peak Trail (strenuous) |
| NOTABLE ANIMALS | California whiptail (*Aspidoscelis tigris*); acorn woodpecker (*Melanerpes formicivorus*; see page 59); big brown bat (*Eptesicus fuscus*) |
| COMMON PLANTS | Blue oak (*Quercus douglasii*); California polypody (*Polypodium californicum*); white pitcher sage (*Lepechinia calycina*) |
| ICONIC EXPERIENCE | Hiking one of the talus caves |
| IT'S WORTH NOTING | PINN is an official recovery site for the California condor, where park biologists manage and monitor dozens of these rare birds. |

**WHAT ARE PINNACLES?** Pinnacles are volcanic rock spires, and the namesake of one of the newest parks in the system. Tiny PINN packs a big punch just a couple hours away from California's bustling Bay Area. The park's iconic rock formations are the result of volcanic eruptions that occurred about 23 million years ago and have made PINN a popular rock-climbing destination. If you're not a climber hoping to get rocked, rest assured that there are plenty of hiking trails where you can see abundant plant life in the chaparral ecosystem, and wildflowers throughout the spring, summer, and fall. Look up, too, and you might see a rare California condor soaring overhead. For maximum comfort, be sure to come in the cooler months—temperatures and conditions are extreme in the summer!

## WILDLIFE

### THE SECRET LIVES OF WILD BEES

**BEES, NO MATTER HOW YOU** feel about them, their importance to the food supply of the planet can't be overstated. Possibly the most well-known pollinators, more than twenty thousand species of bees around the globe join other insects, birds, and even mammals as major players in keeping our food supply functioning. Picture a bee in your mind—you're probably thinking of a plump, hairy bumblebee or a more sleek, still kinda hairy honeybee. Both are common in the spring and summer and eager to bop from flower to flower, gathering the nectar and pollen they need—and while they're at it, providing the pollinating we all depend on. In PINN, you're likely to see bees; probably more than the two types you might already know.

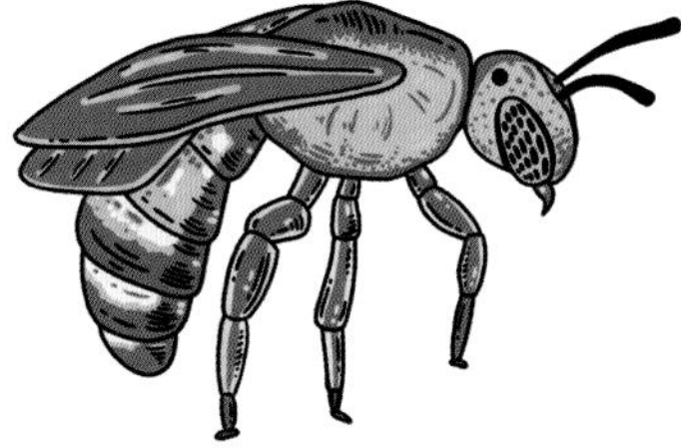

Green sweat bee (*Agapostemon* spp.)

PINN is a small park, coming in at a humble 42 square miles. For comparison, YELL is almost 3,500 square miles. What PINN lacks in size, though, it makes up for in floral and habitat diversity, which, along with its climate, has made it a prime place for bees. In the late 1990s, scientists found nearly four hundred bee species within the then national monument's boundaries. (North America as a whole has around four thousand species of bees.) Since then, more species have been found, and efforts are being made to continue to study these wild native bees. This is especially important because very little is known about wild-bee populations. PINN gives scientists great access to a diverse population of native bees in a relatively small space, conveniently located near heavily farmed land. This provides potential to dig deeper into the secret lives of bees, and it may help ensure that wild bees—which are vulnerable in many areas—can continue to thrive.

Cuckoo bee (*Townsendiella ensifera*)

How can so many bees live in such a small area, anyway? Their lifespans and habitat needs give us some answers. Bees are active for a couple months out of the year, so different species are out at different times. Some bees visit only specific flower species, meaning that various bee species can live in the same area at the same time. Speaking of living, it may shock you to learn that not all bees build hives and have queens (sorry, ladies). Most bee species are solitary, and each female builds her own nest, whether it be on the ground, on stems, or in trees. There are some particularly menacing bees that don't feel that nest-building is a worthwhile way to spend time, though.

About 15 percent of all bee species are kleptoparasites (or cleptoparasites), which means parasitism by theft. Often referred to as "cuckoo bees," they lay their eggs in another species' nest, just like their cuckoo bird comrades. Their eggs hatch earlier than the nest-maker's, and those larvae eat the food that was intended for the original eggs. Those original eggs are either killed by the female cuckoo bee or eaten by the earlier-hatching larvae.

Just as it's possible that you'll see a lot of bees during your time in PINN, it's also likely that you'll witness some fantastic wildflower blooms, thanks to the bees' pollinating efforts. No matter how many bees you see, though, remember to consider that there are many more bees than the honeybee and the bumblebee, and they all play a key role in our global ecosystem.

Interestingly, it's presumed that cuckoo bees aren't great pollinators because they're relatively hairless. They're often confused with wasps for this reason, and they can be difficult to identify without close examination.

Bees can be found all over PINN—more than two hundred fifty species of bees have been spied on Old Pinnacles Trail alone! Look on flowers, on the ground, and in the air for different types of bees. When you see bees, leave them alone and don't provoke them to sting you. If one lands on you, remain calm until it leaves.

## and another thing

## TALUS: A DIFFERENT KIND OF CAVE

**A TRIP TO PINN** isn't complete without a hike through one of its famous talus caves. (See page 255 for more on talus.) These caves aren't the classic caves that most of us picture, though. Instead of forming in the more typical way—where a rock such as limestone is dissolved by water underground—these caves form from rocks and boulders falling from cliffs above into narrow canyons and leaving passageways underneath. These caves are still susceptible to weathering and erosion, and they can be closed suddenly due to flooding. The caves, such as Bear Gulch Cave, are also home to the protected Townsend's big-eared bat (*Corynorhinus townsendii*), so keep your eyes peeled and your voices low when you visit. Bear Gulch Cave is closed from mid-May to mid-July while the bats raise their young.

STATE
ARIZONA

ANNUAL VISITATION
6.4 MILLION

YEAR ESTABLISHED
1919

# grand canyon

## the quintessential canyon

| | |
|---|---|
| SUPERLATIVE | The Grand Canyon showcases the largest section of geologic time visible on Earth |
| CROWD-PLEASING HIKES | Trail of Time (easy); Rim Trail (moderate portions); Bright Angel Trail (strenuous) |
| NOTABLE ANIMALS | Canyon wren (*Catherpes mexicanus*); California condor (*Gymnogyps californianus*); raven (*Corvus corax*; see page 145) |
| COMMON PLANTS | Rubber rabbitbrush (*Ericameria nauseosa*); ponderosa pine (*Pinus ponderosa*); snakeweed (*Gutierrezia sarothrae*) |
| ICONIC EXPERIENCE | Snapping pictures from the Rim Trail |
| IT'S WORTH NOTING | The less-visited North Rim offers tremendous backcountry driving and more secluded viewing opportunities. |

**WHAT IS THE GRAND CANYON?** You can read about its size or how it formed or its constituent rock layers all you'd like, but none of it will matter once you're standing on the rim of the Grand Canyon. The bottom step of the Grand Staircase (the other "steps" are ZION and BRCA) is a place that overwhelms any sense of time or scale. The vast majority of folks head to the South Rim, where there are lodges, museums, and a free shuttle bus. The North Rim is the less-developed, much more wild side of the park. From either rim, the immensity and intricacy of the canyon are a sight to take your scientific breath away.

# WILDLIFE

## TAKING A WALK WITH THE WILDLIFE

**SINCE GRCA IS SO HEAVILY** visited, human/wildlife interaction is frequent. As visitation increases every year, park staff must continually figure out how to manage the relationship between humans and wildlife to keep everyone safe. GRCA's biggest draw is the canyon itself (obviously), so, many people may not even have wildlife on their radar during their visit. It's important to always keep wildlife in mind, though, because you could have an encounter at any time. Three species in the park are prime examples of why humans need to be aware of the impacts of their interactions with wildlife

Huge, beautiful, majestic—no, we aren't talking about the canyon itself, we're talking about elk (*Cervus canadensis*). Bulls (males) weigh up to 800 pounds and have antlers that can reach 5 feet across—they're tough to miss. People often feel compelled to get really close to them to snap a photo or try to pet them. Bad ideas, folks. Elk can be really dangerous (this shouldn't be a surprise) and can very easily injure people, especially during their rut. From late August through October, testosterone runs high, and males bugle really loudly (it sounds like metal rubbing on metal) and seek out cows (females). It's worth noting that elk are invasive to the area.

Raven (*Corvus corax*)

Take a look at the sky in the park—you're likely to see ravens (*Corvus corax*) soaring about. They're very common in the park and are some of the smartest animals out there. Since ravens are so clever, you'll probably see them wreak minor havoc wherever humans enjoy meals outside, especially at campsites. Ravens will do pretty much anything for an easy meal, so if food is left out (even in a bag), they'll get into it. Some ravens have even been seen opening zippers with their beaks. So make sure you securely store your food away as soon as you're done eating!

The most dangerous animal in the park, according to park staff, may shock you. It's the rock squirrel (*Otospermophilus variegatus*), which is common wherever humans congregate in the park. If you don't see one immediately, just wait—they'll find you. Unlike the ravens, though, they'll simply approach humans and beg for food. Often, because they're cute, humans just give them a treat and sometimes let the little critters climb on them. The problem is that squirrels, like all animals in the park, are wild. They are most definitely habituated, but that doesn't mean that they're safe. Squirrels will bite people pretty badly, and they also can carry diseases such as ringworm and plague (yikes).

The good news is that *you* have the power to do something about all of these wildlife problems! Always keep an eye out for wildlife as you're enjoying this (and any) park. Perhaps more important, know that you have some power. Feel free to let humans interacting with wildlife know that it's a bad idea, and that they need to keep their distance to protect themselves and the animals. Also, make sure to let rangers or staff know about any aggressive animal (or human) behavior that you see.

**Elk, ravens, and rock squirrels are all over the park, along with other wildlife. If you hold up your thumb and can cover the entire animal, you're at a safe distance. Depending on the size of the animal, that could be 25 to 100 yards away.**

## BOTANY

### PONDERING THE PONDEROSA

If you're lucky, you may come across a ponderosa stand that feels open and airy. These less-dense stands were the norm before Europeans altered fire regimes in the West. When you see those stands, you're getting a glimpse of these forests as they were before European arrival.

**LOOK AROUND ALMOST ANYWHERE YOU'RE** standing on the North or South Rim and you're likely to be in visual range of at least one ponderosa pine (*Pinus ponderosa*). These stately trees can claim several superlatives, including longest needle of any pine and (more subjectively) the tree whose trunk you most likely want to mush your face against, due to their trademark delicious smell.

Identifying them is pretty simple: Mature trees have a very distinctive thick plating of orange-red bark that stands out in almost any forest, even from a distance. "Mature" is a bit of an understatement, by the way. Ponderosas can live up to 600 years, so anything younger than about 150 years is still an adolescent. And much like human teenagers, young ponderosas look different from their older counterparts. To find young ponderosas, look for darker, gray-black bark and a pointy pyramid-shaped crown. Older trees have much more rounded crowns. The oldest trees have flat-topped crowns—if you need advice, they're definitely the ones you want to ask.

It's hard to miss the orange bark and great smells of a ponderosa pine.

As you travel through the park, keep an eye out for stumps of trees that have been cut down. (It should go without saying that the NPS is not out here wantonly slaughtering trees and laughing maniacally about it. Sometimes trees become damaged or diseased and must be removed for safety reasons.) If you come across such a stump, take a moment to count the rings. Trees lay down two different rings per year—a lighter one in spring, and a darker one for later growth—but it's almost impossible to tell the difference in dead trees, so just divide the number of rings by two to get a really good estimate for that tree's age at death.

If you stay alert, you stand a very good chance of seeing a ponderosa that has been struck by lightning. Lightning strikes *a lot* in GRCA—more than 25,000 times per year on the South Rim alone. The tall ponderosas attract quite a bit of it, but most live to tell the tale! Look for trees that have long scars of missing bark. When lightning strikes a ponderosa, it's conducted through the xylem—part of the tree's vascular system that transports water and nutrients from the roots to the needles. The xylem essentially flash boils and explodes the bark outward. It might look painful, but that action goes a long way toward keeping extreme heat and energy away from parts of the tree where it could do catastrophic damage.

Pretty cool. But the best part is that with ponderosas you can have your tree and smell it too! Go on, get close and get your nose between the plates of some of that orange bark and have a good long sniff. (Take a look before you do, though—lots of insects and spiders live on that bark!) Vanilla, butterscotch, fresh cookies . . . people smell slightly different things, but all of them are wonderfully pleasant. No one really knows what chemical makes the trees smell like that or what its function is, but everyone agrees that the nose knows best in these forests.

**Ponderosas are common on both rims of the canyon in forests that range in elevation from 6,500 to 8,200 feet. The North Rim and surrounding Kaibab National Forest contain some of the oldest ponderosas in Arizona.**

# GEOLOGY

## FOR CRINOID OUT LOUD

When you find any of these fossil treasures, take pictures (including a familiar item for scale), make a sketch, or do an interpretive dance. *Never* move or remove them.

**ONE ESSENTIAL TO BRING WHEN** hiking in GRCA is water. In this arid desert environment, water is scarce and dehydration is an extreme threat. But that wasn't always so. About 270 million years ago, a vast, shallow inland sea covered this area, which was on the western margin of the supercontinent Pangea. It wasn't the first sea to splash around here, but it would be the final one (at least, so far). The remains of that sea and its inhabitants make up the fossil playground known as Kaibab Limestone—the top layer of the Grand Canyon.

Whether or not you are intrepid enough to hike from the canyon's rim to the river below, the Kaibab Formation offers a wonderful chance for you to take a little prehistoric beach vacation. Like any good walk on the beach, even a short hike along the rim will reveal a menagerie of sea life.

Among the creatures in that ancient sea were some that looked more like plants than animals: *crinoids*. At their base was a holdfast that resembled roots and helped them stay in place or move around as they pleased. At the top was a cup-shaped head that looked like a flower. And in between was a body made up of stacked discs with a hole in the center. Those beadlike discs (about the size of a dime) are the most common bits of fossil crinoid you'll find. After the crinoid's death, the discs tended to separate, so look for individual discs scattered about. Lucky fossil hunters will spot columns that are still intact. Crinoids still exist today in the deep ocean, in cold Antarctic waters, and on shallow-water reefs in the Caribbean and the South Pacific.

Sharing the ancient waters with crinoids were two-shelled critters called *brachiopods*. They look like clams, though the two are totally unrelated. Only a smattering of brachiopod species are alive today, but finding their fossils here is simply a matter of squinting at the rocks long enough. Look for, well, shells. Brachiopod fossils look very much like someone has gone around pressing shell halves into bare stone. Look for

# A GRAND SEA

Here are a few of the most common fossils found in GRCA.

## BRACHIOPODS

Brachiopods were prolific in the Permian sea here and lived in dense shoals. Unlike clams, brachiopods have two asymmetrical shells. There are a few types of brachiopod fossils that you can distinguish with a little more work.

### derbyia

Look for fine, tightly packed lines radiating out from the "top" or valve part of the shell (where the two halves would have come together).

### meekella

Look for shells with deep ridges that resemble the ruffles of a curtain or specific brand of potato chip. Cross sections of this shell look like zigzag lines in the rock.

## CRINOIDS

The disc-shaped column segments of these creatures are among the most common fossils in the park. Most are less than an inch in diameter; many are quite a bit smaller. The hole in the center originally held the animal's central nerve.

## SPONGES

Sponges aren't often preserved in the fossil record, so the ones here are a real treat. In life, they were probably no more than a few inches tall. Their fossils look larger than that because of the additional minerals that accumulated on their tissues after death.

fossils that are concave impressions left in the rock, as well as those that look like whole 3-D shells.

Notice any large light-colored discs in the rock, sometimes festooned with intricate branching lines? Those are sea sponges! Just like modern sponges, these ancient ones were colonies of single-celled animals acting like one, larger animal. The whole thing was held together by a structure of silica (natural glass) or calcium carbonate (seashell material) that fell to the bottom and was cemented together after the sponges' death. The resulting rock, called *chert*, is harder than limestone. The sheer abundance of these fossil-sponge-chert nodules has made the Kaibab Formation much more resistant to erosion than it would otherwise be. Thanks, sponges!

Intrepid and experienced hikers will want to check out the Redwall Limestone found in the canyon's walls. Often a series of switchbacks is needed to traverse this 500-foot sheer cliff of red limestone, which is made almost entirely of crunched-up crinoids.

Life flourished in this ancient sea, but all good things must come to an end. The end of the Permian Period, around 250 million years ago, is marked by a mass extinction that wiped out more than 95 percent of species living in water. (The dinosaurs and their extinction have nothing on that number.) Those species that did survive, including relatives of the creatures you see here, went on to rebuild the stunned seas. Today, you can enjoy the glimpse that their fossils provide into an ancient and unfamiliar ocean world.

**The Kaibab Formation makes up the first few hundred feet of rock on the top of the canyon, so almost any trail into the canyon passes through it. Especially noteworthy trails for fossil spotting include the beginning segments of the Hermit and South Kaibab Trails on the South Rim. The North Kaibab Trail is the North Rim's only maintained trail into the canyon. Do your homework and make sure you're prepared for any hike down into the canyon (however short); remember, your return trip will be all uphill.**

STATE
COLORADO

ANNUAL VISITATION
4.6 MILLION

YEAR ESTABLISHED
1915

# rocky mountain

## iconic peaks with hugely popular hikes

| | |
|---|---|
| SUPERLATIVE | The Alpine Visitor Center (11,796 feet) is the highest-elevation visitor center in any national park |
| CROWD-PLEASING HIKES | Bear Lake Trail (easy); Deer Mountain Hike (moderate); Flattop Mountain (strenuous) |
| NOTABLE ANIMALS | Black bear (*Ursus americanus*; see page 252); North American elk (*Cervus canadensis*); western tanager (*Piranga ludoviciana*) |
| COMMON PLANTS | Limber pine (*Pinus flexilis*); Colorado columbine (*Aquilegia caerulea*); Rocky Mountain maple (*Acer glabrum*) |
| ICONIC EXPERIENCE | Driving Trail Ridge Road |
| IT'S WORTH NOTING | Most of the park sits above 7,000 feet—make sure you're acclimated before attempting any physical activity. |

**WHAT IS ROCKY MOUNTAIN?** The Rocky Mountains are a chain of peaks stretching roughly from Alaska to Mexico. ROMO protects just one small portion of the massive mountain chain in Colorado, including the dramatic Front Range on the park's eastern side and the descriptively named Never Summer Mountains in the west. Fertile pine forests and glassy lakes grace the lower elevations, giving way to higher-altitude alpine tundra. It's some of the most easily accessible tundra in the United States, so soak it in. Pack your patience, too—solitude is a rare treat in this heavily visited park.

## WILDLIFE

### A LITTLE TOO ROCKY MOUNTAIN HIGH

**LET'S START WITH A BOLD** but true assumption: You don't live at high altitude. In fact, there's a good chance that you live somewhere near 550 feet above sea level—at least half of the world's population does. Only 6 percent of us live at a mile up or higher. So it's safe to say that, by and large, humans are a low-elevation species. ROMO, however, is practically a highway to high elevation. The park offers some of the most easily accessible altitude in the lower 48, and with that comes a very real hazard: Altitude sickness.

Air pressure drops as you climb in elevation, and lower pressure means that molecules are spread farther apart. So it's not that there's less oxygen in the air—the air around us is a very stable mixture—just that it's more spread out. Every breath pulls in less oxygen simply because there's less around to be captured. Molecules being more spread out is also why temperature drops as you rise in elevation. Being able to take in less oxygen than you're used to is the root cause of altitude sickness, and it does all sorts of wacky things to your body. It can bring on headaches, nausea, dizziness, shortness of breath, lack of appetite, fatigue, and, the best nondescript symptom of all time, a general feeling of malaise.

Hydration is your best friend at high altitude.

Outside of aggravating preexisting conditions, there is no way to predict who will be most affected by altitude sickness. Being in shape, being young, being healthy—none of the standard protections seem to have much influence on who falls victim. Symptoms can start at almost any altitude, but by 7,000 feet—lower than most trailheads in the park—about 20 percent of people experience at least mild effects. By 10,000 feet that number leaps to 75 percent. That means that at the Alpine Visitor Center, which sits at a bonkers 11,796 feet, most people will experience at least some symptoms of altitude sickness.

There's no way to know ahead of time whether high altitude will affect you, but there are steps you can take to minimize the negative impacts of being slightly oxygen deprived. Drink lots and lots of straight up, old-fashioned water and maybe throw in a sports drink or two for some electrolytes. Hydration is your best friend when gaining in altitude, especially in the very dry air of the high Rockies. Being dehydrated will just make altitude sickness, like most things in life, worse. The other big piece of advice is to ascend slowly—try to spend at least one night at 7,000 feet before going higher or setting off on any kind of hike. Driving straight from Denver International Airport (5,300 feet) to the Alpine Visitor Center is not a recipe for a good time, even if all you have in mind is some quick sightseeing.

The good news, folks, is that there's a very simple treatment if you do start feeling ill: JUST GO DOWN. Descend in elevation—get as low as you can go. Drink water and a sports drink, with whatever you take for a headache, and take a nap. You will almost certainly feel better soon.

**ROMO is a high-altitude playground. Go slow, drink lots of water, and explore!**

## BOTANY

### TWISTED TREES

**TRAIL RIDGE ROAD IS THE** highest road in *any* national park, and it's a must-drive if you visit the park during the summer months. (The road is closed to motor vehicles through the winter due to hazardous weather conditions.) While you're driving, you have the opportunity to cruise up to over 12,000 feet (don't get altitude sick!) and experience ROMO from otherwise difficult-to-reach places.

This huge elevation gain means you'll be whipping through various ecosystems. Meadows, trees, and abundant wildlife are present in the montane (below 9,000 feet) and subalpine (9,000 to 11,000 feet) zones. Once you reach 11,000 feet, you'll notice that the trees begin to disappear and give way to tundra. In this transitional zone, the landscape seems otherworldly, scattered with trees that are stunted and crooked from exposure to high winds, frigid temperatures, and ice. They're called *krummholz*—German for "crooked wood." This can happen wherever the environment is nearly unlivable for trees, often due to latitude or altitude. Krummholz trees are pretty common when the conditions are right—in ROMO, affected trees are typically Engelmann spruce, limber pine, and fir.

If you step outside and feel the cool wind (there can be up to a 20°F difference in temperature between Estes Park and Alpine Visitor Center), you'll understand pretty quickly how harsh conditions can get here. How do the trees survive? First, their seeds find safe havens behind boulders or shrubs, or really anywhere protected from the wind. Then, as the trees grow, branches bend with the direction of the prevailing wind. This sometimes gives the trees' branches the appearance of a flag or flags flying from a flagpole, which is another quick way to identify krummholz. Essentially, the trees change their profile to be as accommodating to the wind as possible.

You'll notice that they are dwarfed compared to their lower-elevation counterparts—that's because the lower they grow, the more protected they are from the wind. They're much older too. Krummholz can be up to twice the age of the same species growing in more moderate environs. Low and slow often means a longer life for these hardy trees. While the trees don't know it, their perseverance helps other life thrive here—small mammals and birds rely on krummholz for food and shelter.

If you get out to look closer at the krummholz, remember that you're in a fragile environment. Since life here takes longer to establish itself and thrive, human disturbance can be exceptionally problematic. Tread lightly and step only on established trails, avoiding mosses, lichens, and other plant life. Notice the cool air and consider what could happen to this ecosystem as our climate continues to warm. The creatures and plants long adapted to life in these rugged, cooler conditions will find areas of suitable habitat shrinking, and life will need to adapt to continue to thrive.

Krummholz like this can be found at around 11,000 feet in the park.

**Perhaps the easiest way to see krummholz is by driving Trail Ridge Road and looking out at the trees from the highest stretch of the road.**

# GEOLOGY

## HEADWATERS OF THE WEST

**AN UNPRETENTIOUS LITTLE STREAM MAKES** its way down the west side of the peaks in ROMO. As it runs through Kawuneeche Valley, its waters are cold and clear; you can hear the telltale sound of babbling water and see smoothed pebbles below the surface. It's pleasant, but you could be forgiven for looking right past it. Nothing about this trickle of water gives away just how iconic, important, and powerful it is.

The west side of ROMO is where the Colorado River is born. From here, it flows almost 1,500 miles before spilling into the Gulf of California. (Or at least it used to. The modern Colorado is so heavily dammed and diverted that virtually none of its water makes it that far anymore.) Along the way, the river and its major tributaries touch seven states six national parks. The Colorado River provides water to more than 35 million people; and in its spare time, it carves the Grand Canyon.

The Colorado starts off, of course, much more humbly. Near hiking trails and next to abandoned mining camps all along the Kawuneeche Valley, it's possible to leap from bank to bank. In some spots, you can even keep a foot on each bank and stand directly over the nascent river behemoth.

You might think that such a powerful and extensive river would leave no doubt about itself in the geologic record. Rocks are gossips who love to tell tales about themselves, but like any good gossip, they are unlikely to give you the full story. As recently as 6 million years ago—about the time our ancestors were becoming bipedal—there's no evidence of a river flowing all the way from the Rockies to the Pacific. But 2.6 million years later, suddenly there's a river in full force. What happened between 6 million and 3.4 million years ago is one page of the geologic record no one has found yet. There may well have been an ancestral Colorado that flowed north, rather than west, though scientists are still busy debating the facts and collecting evidence.

## THE COLORADO GOES WEST

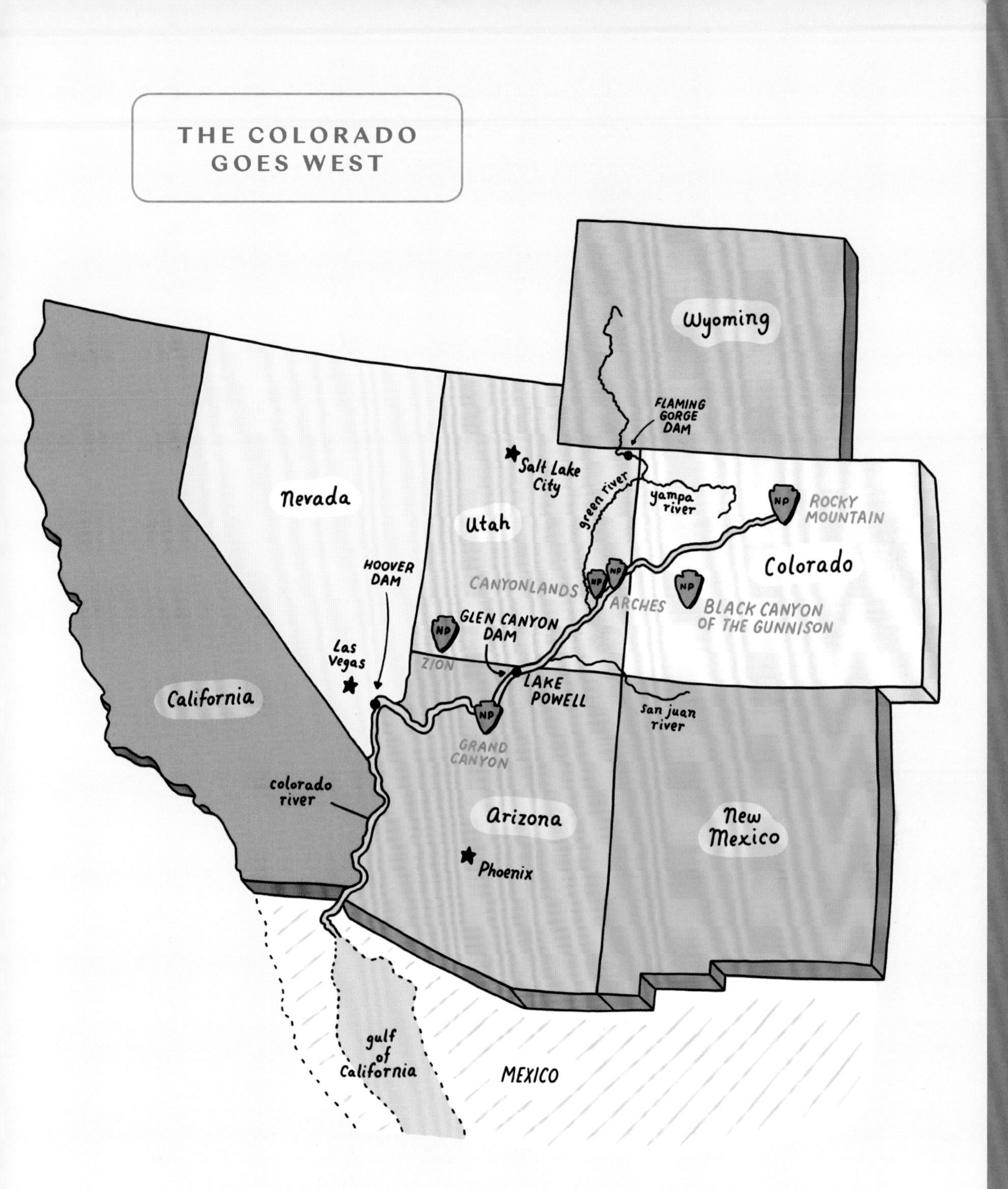

Covering much of the Four Corners region, the Colorado Plateau is a vast geologic province made up of mostly horizontal-lying rocks that have been uplifted high above sea level. Its colorful rock layers, winding rivers, and exquisitely carved canyons (including the Grand Canyon) are home to the densest concentration of NPS sites outside of the Washington, DC area.

How does a river form? Water running downslope cuts into rocks and soil, forming streams that repeatedly merge together and eventually coalesce into a river. The more sediment the water picks up, the faster a river will deepen its channel. The metamorphic rocks that make up the peaks in ROMO don't yield easily to erosion, so the young Colorado River is clear and mostly free of sediment. Downstream, the river flows across the massive sedimentary layers of the Colorado Plateau. As it flows through that giant, uplifted section of crust, the river cuts its way down through 2 billion years of Earth history, picking up so much sediment that its waters become cloudy and colorful.

It's always tempting to look at something as big and mighty as the Colorado River and assume that its story is obvious and its power is inherent. The small, clear stream flowing here gives no indication of what it will become, nor does it yield any secrets about how it came to be. Listen closely, though—this is the sound of a river on its way to greatness.

**The Kawuneeche Valley is on the west side of the park—Grand Lake, not Estes Park. There, you'll find Colorado River Headwaters Trail. La Poudre Pass Trail to Lulu City also features some nice views of the young river.**

STATE
**UTAH**

ANNUAL VISITATION
**4.3 MILLION**

YEAR ESTABLISHED
**1919**

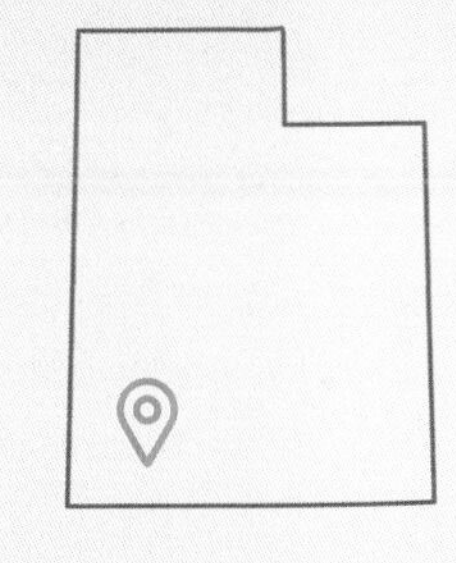

# zion

## colorful cliffs & narrow canyons

| | |
|---|---|
| SUPERLATIVE | ZION is Utah's first national park |
| CROWD-PLEASING HIKES | Weeping Rock Trail (easy); Watchman Trail (moderate); Angel's Landing (strenuous) |
| NOTABLE ANIMALS | Canyon tree frog (*Hyla arenicolor*); peregrine falcon (*Falco peregrinus*); bighorn sheep (*Ovis canadensis*) |
| COMMON PLANTS | Southern maidenhair fern (*Adiantum capillus-veneris*); prickly pear cactus (*Opuntia* spp.); big sagebrush (*Artemisia tridentata*; see page 123) |
| ICONIC EXPERIENCE | Hiking the Narrows |
| IT'S WORTH NOTING | Saturn's moon Titan has slot canyons that are similar to Zion Canyon. |

**WHAT IS ZION?** Zion, a name that often connotes a kind of utopia in various Judeo-Christian religions, is the name Euro-Americans gave to a spectacular canyon in southern Utah. ZION was designated to protect the canyon and its surroundings. This is the middle step of the Grand Staircase—a series of dramatic cliffs and plateaus cut by even more dramatic canyons that stretches from Bryce Canyon south to the Grand Canyon. Here, the green canyon floor is flanked by towering walls of red sandstone—the product of 13 million years of erosion by the Virgin River and its tributaries. During the busy season (March to November), a park shuttle replaces private cars in the most heavily trafficked areas of the park. Hop on and enjoy!

# GEOLOGY

## THE JURASSIC WIND REPORT

**ANY RESPONSIBLE PARK EXPLORER OR** hiker checks the weather before setting off on their adventure. It's a wise and safety-conscious practice, especially in the desert. But by venturing into the heart of ZION, you can come face to face with a wind report from the time of the dinosaurs exquisitely preserved in the towering walls of Navajo Sandstone.

Around 200 million years ago, shifting tectonic plates were causing rivers, lakes, and floodplains in this area to dry out as the climate warmed up. A sea of sand from the northeast slowly engulfed this area, bringing with it conditions nearly identical to the modern Sahara. At its largest, this immense desert covered an area larger than present-day Montana, with dunes tall enough to bury the Statue of Liberty.

Today, that desert is beautifully captured in ZION's walls of Navajo Sandstone. Pay close attention and you'll see three horizontal bands of color in the sandstone: brown on the bottom, pink in the middle, and nearly white on top. You've heard of Neapolitan ice cream (chocolate, strawberry, and vanilla); this is Neapolitan sandstone. No one has figured out exactly why the colors are distributed the way they are (so maybe work on solving that after your vacation). The prevailing thought is that the pink color may most closely represent the authentic and unaltered color of the dunes.

More important than the color is the information coded into this petrified sand sea. It won't take you long to spot one of the most striking features of Navajo Sandstone: alternating bands of diagonal stripes in the rock. Known as *cross-bedding*, these lines are a Jurassic wind report. Sand dunes may seem random, but they form and behave in predictable ways. The wind always picks up the sand from the gently sloping upwind side of a dune and deposits it on the much steeper downwind side. That process is repeated over and over again as dunes rise, fall, and shift across the landscape. Navajo Sandstone, the rock created when those dunes turned to stone, preserves a beautiful and clear history of ancient wind patterns—what you see today as braided and intersecting lines in the canyon walls.

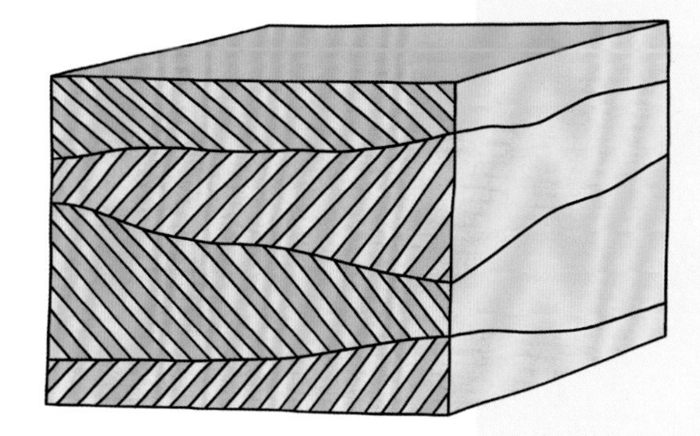

More than 200 million years after its formation, the beautifully cross-bedded layers of Navajo Sandstone preserve a clear snapshot of ancient wind patterns.

Not all was dry in that Jurassic desert, though. At the Canyon Trail Overlook parking lot (as well as along the trail itself), you can see evidence of what was once an oasis in the desert. Look for a bright red streak between the lower brown and middle pink layers of the sandstone. Now, get closer! Within that layer, careful observation should bring mudcracks into focus. Though they are more than 100 million years old, they look just like modern mudcracks. Other evidence of abundance at the oasis is a little more difficult to spot, but nonetheless tantalizing; faint root traces from long-vanished plants mostly resemble ghostly halos in the rock.

There are other places in the world that have preserved ancient dunes and deserts, but none quite like the Navajo Sandstone in ZION. As you take in the stunning scenery, spend some time getting lost in the shapes and patterns of the rock. Can you feel the hot Jurassic wind?

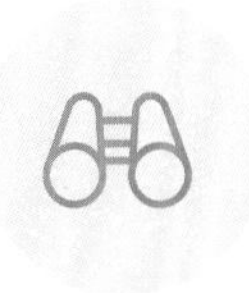

**Huge exposures of Navajo Sandstone make up much of the rock in the park. The east side has some exceptionally clear views of cross-bedding. Canyon Trail Overlook is located just east of Mount Carmel Tunnel along the park's main road. The hike from there is easy and popular.**

# BOTANY

## HANGING IN THE GARDEN

The Zion snail (*Physa zionis*) is the smallest snail in the world—about the size of a pinhead. It's found only in some of ZION's hanging gardens and nowhere else.

**WATER IN A DESERT IS** always sure to attract attention. In ZION, with the exception of the Virgin River, your first clue to its presence might be streaks of green cascading down sheer rock faces. These are the hanging gardens of ZION: uniquely blended communities of plants clinging to seeps and springs that emerge from bare rock.

Hanging gardens are unique to the Colorado River and its tributaries, like the Virgin River. What you see here happens almost nowhere else and requires an essential mix of three ingredients: a steady water source, intrepid plants, and just the right kind of geology. ZION has them all. Navajo Sandstone, which makes up a large portion of the canyon walls, is porous and acts like a stone sponge. Rain and snow seep into it and happily percolate for a very long time. (One study found that some of the water emerging in the gardens was 4,000 years old.) When water meets the less permeable Kayenta Formation beneath the Navajo Sandstone, it is forced sideways—sometimes emerging from the rock face in seeps and springs.

Hanging gardens can be as simple as a single trickle of water supporting a few plants, or they can be wildly complex networks of plants that stretch across entire rock walls. Some even have enough water that pools form at their bases. Look for hanging gardens on sheer rock faces, on debris piles beneath canyon walls, and in alcoves. Water and weak acid (naturally secreted by plant roots) erode the rock and create those alcoves and debris piles—rare opportunities for shade in the desert and ideal habitats for plants. A garden that starts as a relatively simple seep might eventually grow into something much more sprawling.

Each seep supports its own unique and highly diverse mix of plants. Only about a dozen species of plants live in hanging gardens and nowhere else, but the vast majority of residents are plants that can be found elsewhere—usually in far wetter and cooler environments. Some of those characters got

## FERNS, GOLLY

Each hanging garden is unique, but here are some of ZION's more noteworthy garden inhabitants. Never touch the plants or rocks—your hands will irreparably damage these magical places.

**Southern maidenhair fern (*Adiantum capillus-veneris*)** These ferns are among the most common and abundant in ZION's hanging gardens. Look for wiry dark red-brown stems and irregular wedge- and fan-shaped leaves.

**Zion jamesia (*Jamesia americana* var. *zionis*)** Jamesia can be found throughout Utah, but this particular variety exists only in ZION. Look for white, five-petaled flowers when this woody shrub blooms from May through September.

**Golden columbine (*Aquilegia chrysantha*)** These large yellow flowers bloom in late spring and early summer. Keep an eye out for similar flowers with an orangish red color—hybrids between golden columbine and a red-flowered variety. You can often spot all three varieties in the same place.

**Jones reedgrass (*Calamagrostis scopulorum*)** Look for clumps of this native grass growing out of crevices in the rock. Reedgrass stalks are topped by feathery seed heads. It is especially common along the Narrows.

**Cardinal monkey flower (*Mimulus cardinalis*)** Look for these red flowers from May through August. At other times of the year, look for downy, toothed leaves. This flower is native to much of the United States, but ZION is the only place it is found in Utah.

**Tall fescue (*Festuca arundinacea*)** This invasive grass, which can grow more than 3 feet high, was introduced to North America from Europe. In ZION it crowds out native species at the base of the gardens. Be sure to give it a dirty look.

here via long-distance dispersal of seeds or spores; others grew more widely during cooler glacial periods and then became isolated in the gardens as the climate warmed. The modern isolation of the Virgin River has fostered species in ZION that are similar to, but unique from, those in other hanging gardens.

The gardens here are more easily accessible than anywhere else they occur. So strap on some sturdy shoes and set off into the park to see these special places. The gardens are magnificent, but they are no match for human greed or overuse. Never touch the plants or the rock walls; instead, take in these vertical oases with only your eyes. They are a tall, cool drink of green in the land of red rock.

**Keep an eye out for hanging gardens all over the park along trails as well as roads. Some especially noteworthy and easily accessible trails that access hanging gardens include Weeping Rock (which features trailside interpretation about the plants); Lower, Middle, and Upper Emerald Pools; Riverside Walk leading to the Narrows; and of course the Narrows themselves.**

## DO YOU HEAR WHAT I HEAR?

**AMPHIBIANS MAY NOT BE THE** first creatures that come to mind when you think of life in the deserts of southern Utah. And while you're right—deserts aren't usually an ideal place for amphibians to set up shop—there *are* some brave amphibious creatures that enjoy the few moist areas of ZION. Seven species of amphibian live in there, six of them frogs or toads and one of them a salamander.

Wildlife is unpredictable at best, you can never be guaranteed to see a particular animal when you visit a park, and it should be a rare delight to come across one. That's just how animals work—under the best of circumstances, they're afraid of humans. Imagine, for a moment, a different way to engage with wildlife than using your eyes. Every park tells a story with its sounds, and ZION is no different.

There is one particularly vocal little frog, the canyon tree frog (*Hyla arenicolor*), whose males sing mating calls to attract a female from spring through early summer. These males usually start singing after sunset, but sometimes start as early as the afternoon, and their song is actually quite loud for their little bodies (which are about 2 inches long). It sounds a little bit like a bleating goat, but very fast, rhythmical, and amplified. Some have described it as sounding like the frog is in a tin can; others say it sounds like a bird.

These frogs are abundant in the southwestern United States, and while we can't guarantee it (we know, you get it now), it's very possible that you'll see one in the park if you keep your eyes open and know where to look. Canyon tree frogs are typically on rocks at any stream, and on rock ledges, where they use their large, suction-cup toes to cling and hang out after spending short spells in the water. Their coloration

Canyon tree frog
(*Hyla arenicolor*)

varies based on environment, but here you're looking for tan, gray, or olive on their toadlike, warty bodies. You might also see a cluster of frogs in rock crevices.

Canyon tree frogs are also the focus of some ongoing research on chytrid fungus. This terrible fungus has decimated about one-third of the *world's* amphibian population, so when it was initially found in ZION in these frogs, it raised some concerns, especially since ZION is a relatively isolated area. However, the research indicates that the fungus doesn't have an effect on population numbers at all. Scientists don't yet know exactly why the fungus isn't killing off the frogs, although one guess is that their low-water lifestyle and minimal reliance on breathing through their skin helps them evade fungal devastation. As always, if you see a canyon tree frog (or any other wildlife in the park), don't handle it; instead, have a good long look and listen.

**Keep your ears open, because you'll probably hear the frogs before you see them. They're common wherever there is water, including the Emerald Pools and Hidden Canyon.**

STATES
# WYOMING, MONTANA, AND IDAHO

ANNUAL VISITATION
**4.1 MILLION**

YEAR ESTABLISHED
**1872**

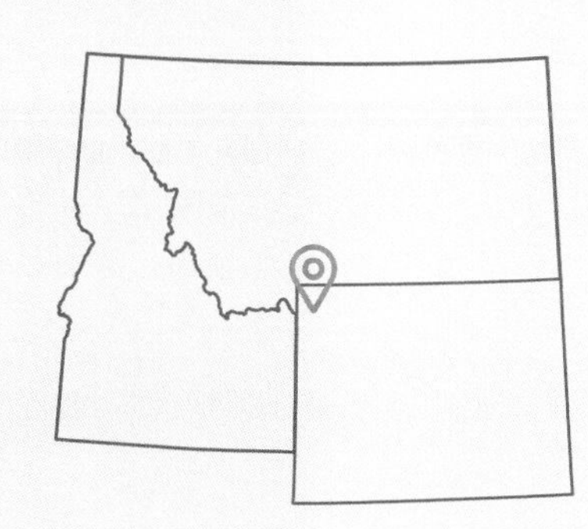

# yellowstone

## a giant, geyser-filled caldera ringed by mountains

| | |
|---|---|
| SUPERLATIVE | YELL is the first national park |
| CROWD-PLEASING HIKES | West Thumb Geyser Basin Trail (easy); Fairy Falls Trail (moderate); Mount Washburn Spur Trail (strenuous) |
| NOTABLE ANIMALS | Bison (*Bison bison*); bald eagles (*Haliaeetus leucocephalus*); coyotes (*Canis latrans*) |
| COMMON PLANTS | Lodgepole pine (*Pinus contorta*); big sagebrush (*Artemisia tridentata*; see page 123); prickly pear cactus (*Opuntia polyacantha*) |
| ICONIC EXPERIENCE | Watching Old Faithful erupt |
| IT'S WORTH NOTING | Half of all the hydrothermal features in the world are found in YELL. |

**WHAT IS YELLOWSTONE?** YELL is named for the Yellowstone River, which flows through it. But the big storyline here is a really big volcano. The park sits directly on top of one of the largest active volcanoes in the world, a hotspot of immense proportions that fuels more than 10,000 hydrothermal features aboveground. Most of the developed park is within a caldera roughly the size of Rhode Island, formed after the hot spot's last major eruption. Ringed by mountains, the park is home to an abundance of wildlife, sculpted glacial features, and chilly weather—even in the summer.

# GEOLOGY

## E.T. PHONE YELLOW-STONE

**IF YOU IMAGINE THE SCENE** on the surface of an alien planet, you might conjure something that looks like YELL's geyser basins—scorched soils, strange colors, nose-wrinkling smells, bubbling mudpots, and steam rising from the ground itself.

YELL's hydrothermal features are powered by the commingling of moisture from the mountains above with the immense heat generated by the magma chamber belowground. Changes in the behavior of the features provide clues to the complex and ever-shifting plumbing beneath the surface. As you explore the park, keep an eye out for still-standing dead trees with a white ring around their bases. Known as *bobby socks*, these rings are a sign the trees were killed by shifting thermal waters underground. As the trees absorb silica (natural glass) from the thermal water, they are gradually "frozen" in place and won't burn or fall down.

But thermal waters don't just harm life; they also enable it. Aboveground, these waters host an astonishing diversity of creatures, known collectively as *extremophiles*, which thrive in some of Earth's most extreme environments. In Norris Geyser Basin, microbes flourish in water with the pH of stomach acid; near Old Faithful, microbes thrive in water as basic as baking soda. There are pools with temperatures well above boiling, and slightly cooler (but still hot!) runoff streams. Some water is loaded with arsenic; some is brimming with mercury. Life exists in every single one. This kind of extreme environmental diversity makes YELL a living laboratory for studying our universe. And in this lab, you don't even need a microscope to see the microbes—just look for bright colors.

Until the 1960s, no one knew that those colors were actually microbes—not just surviving, but actively thriving in the thermal waters. Then an enterprising park biologist, often armed with little more than a tin can and a thermometer, proved the existence of not just one kind of life, but scores of it.

## EXTRATERRESTRIALS OF NORRIS

The bright colors and distinct smell of YELL's microbes easily alert the savvy traveler (i.e., *you*) to their presence. Color—correlated to location—is all you need for a cursory ID most of the time. Use this handy guide at Norris Geyser Basin. Entire ecosystems are based in these waters, so look for flies and spiders too.

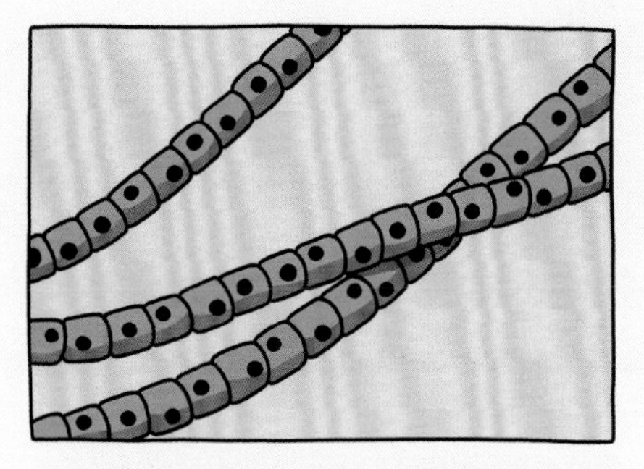

***Zygogonium*** **(algae, 68° to 96°F)** This algae can't take the high heat and is often found in cooler water next to green (and very heat-tolerant) *Cyanidioschyzon*. These mats provide a habitat for acid-tolerant ephydrid fly larvae as well.

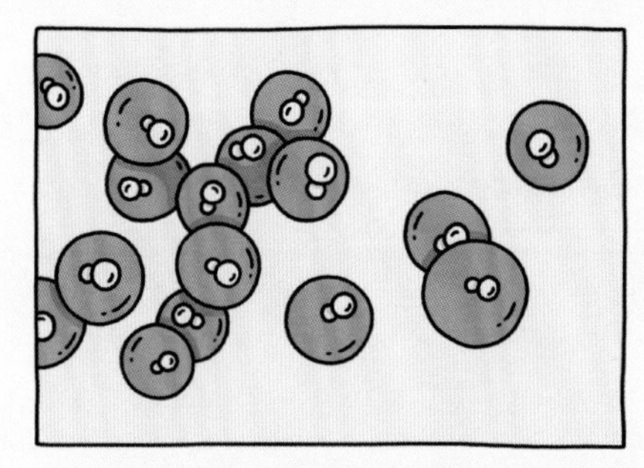

***Cyanidioschyzon*** **(algae, 100° to 126°F)** This is one of the most heat- and acid-tolerant algae known.

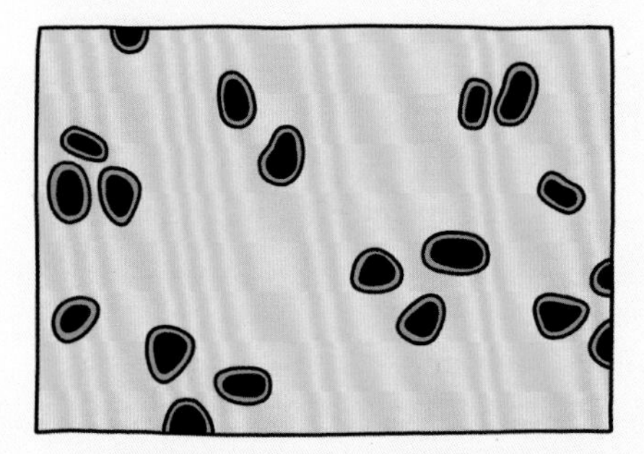

***Metallosphaera*** **(archaea, 122° to 176°F)** This archaea can live in places even other extremophiles can't. It's named for its ability to dissolve metal.

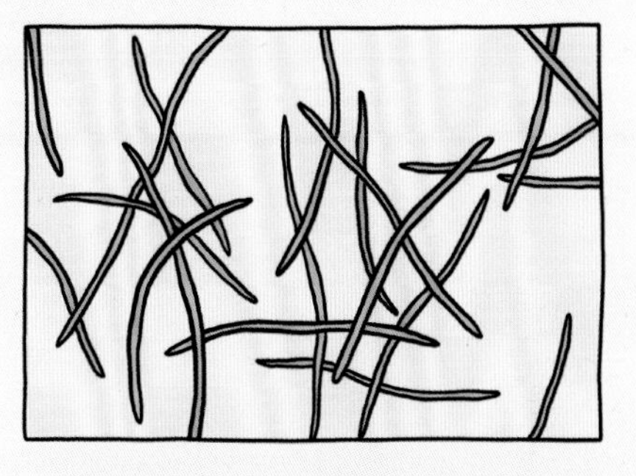

***Hydrogenobaculum*** **(bacteria, 131° to 162°F)** Look for thin yellow and white strands streaming in the water. These bacteria are responsible for the rotten-egg smell—they live on hydrogen sulfide. In the absence of that, they eat arsenic.

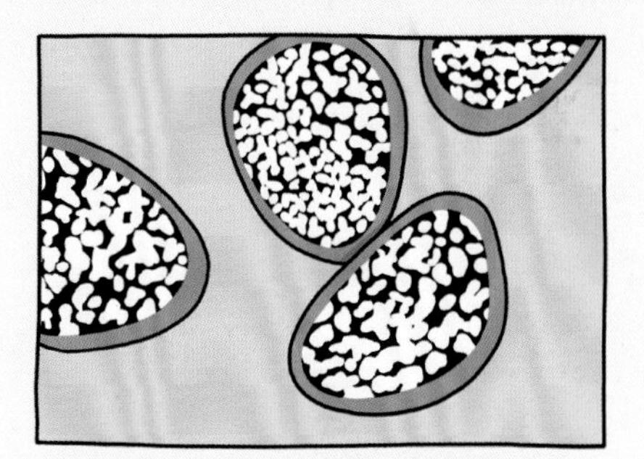

***Sulfolobus*** **(archaea, 149° to 188°F)** Also responsible for the rotten-egg smell, this microbe was one of the first found in Norris and is one of the first archaea ever discovered. Nowadays, scientists are studying viruses that have been found to live on these critters.

Algae and bacteria were there. As were archaea—single-celled organisms believed to be among the first life on Earth, more distinct from bacteria than plants are from animals. Archaea might well be the most abundant life on Earth, though most have proven impossible to culture in a lab and remain difficult to study.

Since the 1960s, scientists have found extremophiles in waste piles at mines, around vents on the ocean floor, and many other places. The extremophiles here remain special not only for their ease of access but also for their wondrous diversity. Astrobiologists and planetary geologists flock to YELL because, if we can understand how life ticks along in these extreme environments, we can learn to look for it in the many extreme environments that exist on planets elsewhere in our cosmic neighborhood. Past or present evidence showing we are not alone will probably look a lot more like these microbes than other larger, more familiar life forms. YELL's extremophiles ripped up the playbook on the "limits of life" and sang the news from the highest peaks: Move over, little green men; extremophiles are the new extraterrestrials.

**These extremophiles can be found in Norris, Upper, Midway, and Lower Geyser Basins. Never leave the boardwalks or throw anything into thermal features. Interpretive signage near some features explores the link between extremophiles and the search for extraterrestrial life.**

## BOTANY

### LODGEPOLES LOVE LAVA

**WITHIN YELL, YOU'RE NEVER TOO** far from a lodgepole pine (*Pinus contorta*). Look up and look around because these tall, skinny conifers are everywhere. They make up the vast majority of trees in the park, often growing in dense stands like some kind of stately supermodel clique.

The reason for their dominance is the same reason for pretty much everything about YELL—the churning dome of magma gurgling just below the surface. During eruptions, that magma comes into contact with and melts some of the earth's crust. This special mix of minerals creates a rock called *rhyolite*. After dynamic and fiery beginnings, rhyolite retires into its rocky form, in which it weathers into an acidic, nutrient-poor soil that most plants avoid. But lodgepoles love it.

Take a look at the plants growing in the understory of a lodgepole stand. What plants, you say? Exactly. That rhyolitic soil doesn't support much besides the lodgepoles. Notice the roots of downed trees. For their height, lodgepoles have almost comically shallow roots. This is an economical choice—water drains out of these volcanic soils very rapidly, sinking out of the reach of plants too quickly to make deep root systems worthwhile. Take a nice long gander at those tall trunks. They are more or less naked up to the canopy, because lodgepoles are totally shade intolerant. If some obstacle, like another tree, begins to shade a lodgepole branch, that branch will die. Give them sun or give them death.

As you explore the park, keep this in mind: Where you see lodgepoles, you are seeing rhyolite—cooled magma. Those sprawling forests of lodgepoles are the signposts of immense

Massive amounts of lava from past eruptions have given rise to towering stands of lodgepole pines throughout much of YELL.

Squirrels and chipmunks dine on lodgepole seeds; grouse eat the needles. Larger animals, like elk, deer, and bears, find shelter and shade under the canopy.

lava flows, cataclysmic eruptions demarcated with trees. It's fitting that lodgepoles have their origins in something as explosive and fiery as rhyolite. Many lodgepoles also need fire to reproduce—some trees produce cones sealed with resin that melts only in the high heat of forest fires. Not all lodgepoles make these cones, but the ones that do tend to live at lower elevations where fire is more frequent than higher elevations.

The cozy relationship between lodgepoles and the YELL supervolcano does have some drawbacks. Sometimes tree roots tap in to thermal waters laden with silica, a natural glass that slowly kills the trees. All that's left are the dead trunks with a white ring around the base. The silica in those trees means that they don't burn in fires and they don't fall down; instead, they become a small window into YELL's complex and ever-shifting underground plumbing.

The absolute dominance of lodgepoles in YELL can start to seem monotonous after a while, but not if you keep in mind that those svelte coniferous icons are a bright beacon that screams out *things are quiet here now, but you are standing in a forest grown from the cooled remnants of catastrophe*. Happy lava hunting!

Lodgepoles are common throughout the park.

## WILDLIFE

### STAYING COMFORTABLE COMES WITH COSTS

**YOU MAY KNOW THE BASICS** of the story—there were tens of millions of bison (*Bison bison*) roaming North America up until the mid-1800s, when they were systematically hunted to near extinction as a means of subduing Indigenous peoples. By 1900, there were only twenty-three wild bison left in the United States, all in YELL. After extraordinary conservation efforts, bison numbers recovered somewhat, and YELL continues to boast the largest wild population living on public land.

The bison in YELL haven't left the area since their near-extinction, and 2,500 to 5,000 live within its boundaries during any given year. There are two distinct bison herds in the park: the central herd (which breeds in Hayden Valley) and the northern herd (which breeds in Lamar Valley and surrounding plateaus). Some of the bison in the central herd choose to spend their winters in the more comfortable areas of the park, where thermal features have warmed the ground and reduced or even eliminated the winter snowpack entirely. Bison who feed where the snowpack is thick use their massive heads to move snow away to get at the plants underneath.

The iconic male bison, often seen as a symbol of the NPS.

Look for hoofprints and scat near thermal features, which are evidence that bison have been in the area. The plants that grow here (mostly grasses and sedges) are bison food. These plants aren't abundant, but scarcity isn't the worst problem—because they grow so close to the thermal waters, they absorb high concentrations of fluoride and silica (natural glass). Both substances occur naturally in plants, but much higher levels are found in plants near thermal features. A little bit of fluoride is a good thing, but digesting large amounts of it might prevent adolescent bison from forming strong teeth. And large amounts of abrasive silica increase the damage by wearing down bison teeth quickly. These factors are a terrible combination for the thermal-loving ungulates.

Teeth may seem like an insignificant sacrifice in the greater scheme of things, but bison who winter in the thermal areas live shorter lives than their counterparts in other areas of the park. Some scientists suggest that they lose 5 to 8 years of life because of their diet's effect on their teeth. Given that the average bison lives 14 to 16 years, that's a significant loss! However, wolf populations are denser in this part of the park, potentially leading to more predation of bison, and the winters are much harsher. All of these factors contribute to a shorter lifespan and higher mortality rates, and it's impossible to tell which has the biggest impact on bison mortality.

Life in YELL is complex, and living on a supervolcano has perks and drawbacks. Sure, it's easier to access food in the more comfortable winter respites, but is it worth potentially losing years of life? Next time you get stuck in a "bison jam" in the park, feel free to spark a lively "would you rather" discussion in the car while you wait for the animals to pass.

**Bison are found throughout the central and northern regions of the park year-round. Large concentrations of them gather in the Hayden and Lamar Valleys during the breeding rut of July through August. DO NOT try to look at their teeth (or touch them at all).**

STATE
WYOMING

ANNUAL VISITATION
3.5 MILLION

YEAR ESTABLISHED
1929

# grand teton

## mountains, hold the foothills

| | |
|---|---|
| SUPERLATIVE | The Tetons are the youngest mountains in the Rockies |
| CROWD-PLEASING HIKES | Lakeshore Trail (easy); Phelps Lake Trail Loop (moderate); Garnet Canyon (strenuous) |
| NOTABLE ANIMALS | Pronghorn (*Antilocapra americana*); bison (*Bison bison*); bald eagle (*Haliaeetus leucocephalus*) |
| COMMON PLANTS | Engelmann spruce (*Picea engelmannii*); big sagebrush (*Artemisia tridentata*); bracken fern (*Pteridium* spp.) |
| ICONIC EXPERIENCE | Watching the sun set behind the peaks |
| IT'S WORTH NOTING | There are plenty of hikes for all abilities that lead to stunning glacial lakes. |

**WHAT IS GRAND TETON?** Grand Teton is the central peak in the Teton Range, which runs just 40 miles from north to south. GRTE encompasses the entirety of the range and much of the valley floor of Jackson Hole. Climbers come to summit the nearly 14,000-foot peaks—at night, look for their headlamps flashing along mountainsides in the darkness. Park roads offer stunning panoramas and unbelievable sunset vistas. The range's modern name comes from some apparently very lonely French fur trappers who thought the jagged peaks resembled nipples. Anatomical inaccuracies aside, the Tetons are an iconic part of the West that will stop you in your tracks no matter how many photos you've seen.

# GEOLOGY

## MOUNTAIN-BUILDING STRETCH MARKS

**MOUNTAINS SEEM LIKE PERMANENT THINGS** to us, so you might not think of the Tetons as looking young. In the United States, there are mountain ranges that are super-old (hello, 300-million-year-old Appalachians) and then there are the Tetons, the babies of the bunch. They come in at a measly 10 million years or so and are in fact still rising. You can even see it happen!

These mountains exist because forces deep underground are stretching this part of the North American tectonic plate in response to events that occurred long ago. Somewhere around 70 million years ago, the Pacific Plate began to work its way underneath, or subduct, the North American Plate. Instead of diving deep down beneath the continental rock like most oceanic plates do, the Pacific Plate stayed dang near horizontal as it pushed under the continent. Think of the Pacific Plate like a spatula and the North American Plate like a pancake—sliding the spatula underneath makes the pancake bulge upward, just like the mountains in the western United States.

The stress of the Pacific Plate moving steadily northeastward from the western margin of the continent caused this part of Earth's crust to bulge upward. In some places, the crust stretched past its breaking point, creating new faults (cracks in the bedrock) and reenergizing old ones. Huge blocks of the rock slipped past each other along those faults, like champagne corks under pressure. One edge of the block of rock shot skyward (the Tetons), while the other hinged downward (Jackson Hole). Mountains built by the colliding of plates, like the Himalayas, often have foothills around them. The stretching going on here sends mountains straight up; no foothills needed.

Get yourself to the floor of Jackson Hole and sweep your eyes across the spectacular eastern front of the Tetons—you are staring directly at one of those champagne-cork faults! The mountains are a block of Earth's crust tilted up into the

If you drive into the park from the west, you'll see the much gentler western face of the mountains—the part of the fault block that is dipping down. It takes more than 10 miles for the western front to drop the same thousands of feet that the eastern front plunges in just a few miles.

## ONE HUGE FAULT

The eastern front of the Teton Range is one enormous fault scarp. Regional stretching is pulling the crust apart, causing a huge chunk of rock (the Tetons) to push skyward.

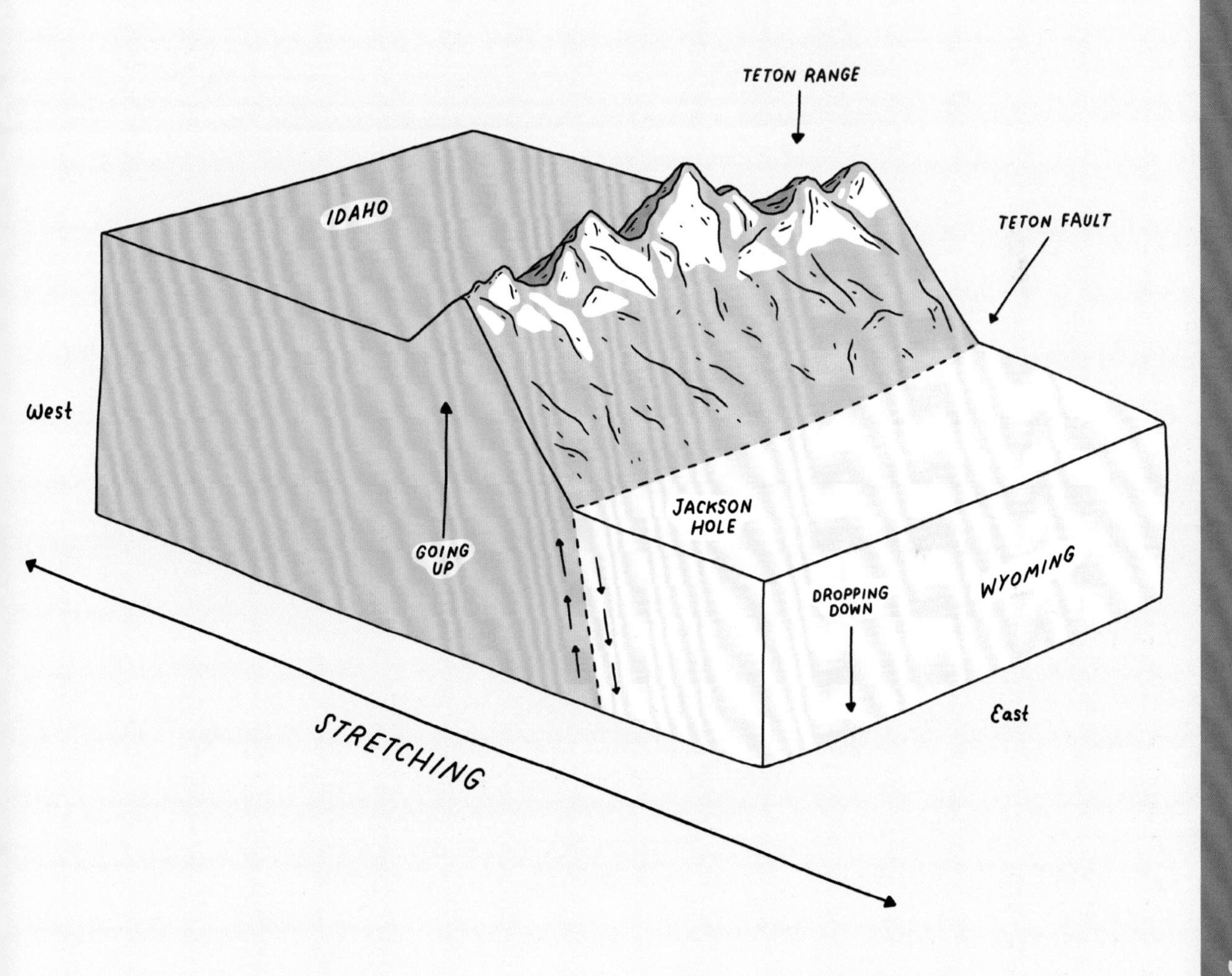

sky. It's happened in only the last 10 million years or so, but this fault has been busy; thousands of earthquakes have nudged this fault block progressively higher, until a few feet of movement have become a few thousand feet.

Glaciers erased most of the evidence of the Tetons' initial ascent, but quakes since the end of the Ice Age have left scars that you can still see. Fault scarps—think of them as the stretch marks of mountain building—can be difficult to recognize at first. Let's start with an easy one near String Lake. Just across the water at the base of the peak, look for an incongruous streak that looks like a giant has run their finger along a section of the mountain. That ledge is a fault scarp more than 100 feet high—a scar generated by a series of quakes over just the past 14,000 years as large coherent blocks of rocks slid past each other.

Now that you've seen one, head out to explore and look for more! The base of the range, aka the length of the fault, is littered with scarps. Some are small, only 30 feet or so high; others are as big as or bigger than their String Lake counterpart. Shadows formed by the steep banks and ledges of the scarps and strange tree-growth patterns are often good signals that a scarp is lurking nearby.

Small quakes happen with frequency along the Teton fault, though it has been about 5,000 years since that baby really let loose. It's impossible to say when the next "big one" is coming, but now you know better than to let the seeming quiet fool you. These mountains are growing, and the evidence is all around you.

**Fault scarps are all over the base of the Tetons' eastern face! The easiest and most recognizable can be found at the base of Rockchuck Peak. To get there, simply follow Jenny Lake Scenic Loop to Cathedral Group Turnout.**

**IT'S SAFE TO ASSUME THAT** you haven't come to GRTE with the intention of seeing anything like a giraffe or traveling back through tens of thousands of years of history, but if you spend enough time in the valley next to those towering peaks that is exactly what you will see and do.

Pronghorn (*Antilocapra americana*) are known by a lot of different names. In Wyoming—home to more pronghorn than anywhere else on the continent—many call them "antelope," but the truth is so much cooler. Pronghorn are the sole North American survivor of a once diverse family that has just two other living branches: giraffes and okapis. No one knows exactly how they came to be, but by the dawn of the Ice Age some 2 million years ago, pronghorn were firmly established in North America.

Ice Age North America was a place literally roamed by lions, tigers, and bears (oh my), not to mention big cats of the saber-toothed and cheetah-adjacent varieties. The pronghorn is the only one to outlast them all, but you don't just forget the experience of being hunted by enormous, swift predators. Their Ice Age origins equipped pronghorn with phenomenal speed, which they retain today. They can run at sustained velocities of 45 to 50 miles per hour. That makes them almost as fast as a cheetah, but with much greater stamina. In a single bound, a pronghorn can cover more than three times its body length. Grab some binoculars and (if you can get close enough for a clear view) take a look at their hooves—two long, pointed toes taper delicately outward, built for grip and speed; unseen are cushioned undersides that absorb the impact of all that running.

# WILDLIFE

## SURVIVOR—ICE AGE AMERICA EDITION

Pronghorn are an iconic resident of the Great Plains and western United States.

Every fall, the pronghorn in the park migrate to more favorable winter grounds. Despite their speed and agility, pronghorn don't really jump, so the encroachment of humans and our fences is problematic for them.

Like pronghorn elsewhere, the several hundred animals that spend their summers in the park hang out in a mixed-gender herd. You can generally identify males by their characteristic facial markings: a black mask sweeping back from the nose and black patches on the jaw below their eyes. Very classy.

Both males and females sport the species' namesake pronged horns, though the males' are typically larger. Take a nice long look at that flashy headgear; it's a true oddity in the natural world. What grows on these animals' heads aren't actually horns or antlers, but a unique blend of both. Like horns, the pronghorns on a pronghorn have a core of bone, covered with a keratin sheath. Like antlers and unlike horns, pronghorns have branches (prongs, if you will) and are shed every year. No other animal has anything like it. Pronghorn grow their head ornaments all year; you'll see the largest ones in August and September.

See any cute baby pronghorn? Females usually give birth to twin fawns in the spring. The fact that they can easily outrun you after only a few days is the least badass thing about them. Those adorable fawns just spent part of their time in utero battling it out with a bunch of other embryos. They twist, pull, and jab at each other with aptly termed "necrotic tips" until there are just two victors left to be birthed. Pronghorn are the only mammal known to science with a gladiator ring for a womb.

These stately, swift Ice Age giraffe cousins are cute, athletic, and just the right amount of vicious—an animal almost anyone could love.

**Pronghorn favor open environments with low vegetation and are most active at dusk and dawn. Look for them in sagebrush meadows throughout the valley floor of Jackson Hole. Teton Park Road usually provides some views, as does the aptly named Antelope Flats section along Highway 191/26/89.**

## BOTANY

### A BRUSH WITH THE SAGE

**BITTER. A TOUCH OF ROSEMARY** and Vicks VapoRub. Christmas. Berries. Evergreens. Those are all ways to describe the smell of one of the West's most iconic plants: sagebrush. Stick your nose in and have a whiff! This low, silvery-green shrub has been used and respected for thousands of years by the original inhabitants of this continent. It has befuddled ranchers and captured the hearts of poets. But of course sagebrush doesn't care about any of that; it's out here growing its best, grabbing moisture, and throwing shade—all of which make its ecosystem among the most diverse in the region.

Sagebrush is related to a host of awesome plants—like wormwood (used in absinthe) and tarragon—but not to the sage you cook with (that's a *Salvia*). This tough shrub has dominated the valley floor of Jackson Hole since . . . there was a hole. The vast majority of it is big sagebrush (*Artemisia tridentata*). In the early spring and summer, you can spot the three-lobed leaves that inspired its scientific name. Those relatively big leaves later wither and die in the summer heat, but that doesn't stop this hardy plant. Big sagebrush also grows a second set of smaller, more irregularly shaped leaves that carry on the work of photosynthesis not only during the height of summer but all the way through the depths of winter. While there's still snow on the ground, sagebrush is photosynthesizing away, getting a leaf up on the competition.

Big sagebrush is a go-getter that smells great and can perform photosynthesis all year long.

The wiry stalks you might see sticking up from the rest of this shrub's crown are special branches that handle all the plant's reproductive needs. If you're in the park in late summer or fall, you may glimpse little yellow flowers on the bushes. Sagebrush is wind-pollinated, so there's no need

The silvery sheen of sagebrush is from tiny hairs on the leaves. They create a barrier that helps moderate leaf temperature.

for flashy flowers. Instead, when it's time to reproduce, those diminutive flowers release tiny seeds and big willowy plumes of pollen. Pass the Benadryl, please.

Sagebrush steppes may look like monocultures, but they are incredibly diverse, with more species of plants than any other environment in GRTE, save for wetlands. You probably won't have to look too long before spotting some pronghorn; these flashy ungulates are practically synonymous with sagebrush meadows. They survive the harsh winters here on a diet of mostly sagebrush leaves and stems. If you're lucky, you may spot moose and deer that come to browse the small, protein-rich seeds.

Keep an eye out for the aptly named greater sage-grouse (*Centrocercus urophasianus*) and its very threatened cousin, the Columbian sharp-tailed grouse (*Tympanuchus phasianellus columbianus*). Listen for the rhythmic sounds of sagebrush crickets (*Cyphoderris strepitans*). Those nightly calls are mostly from uncoupled males. After coupling, some females eat the males' hind legs for nutrients, so it's probably safe to assume that coupled males have lost their voices.

Sagebrush may be iconic, but like pretty much every other living thing it is threatened by loss of habitat and climate change. By some accounts, at least half of all the sagebrush in the United States has disappeared since European settlement. Spend time in GRTE's protected meadows, though, and you'll quickly understand that sagebrush flats are feasts for the senses that should not be missed.

**Much of the valley floor of Jackson Hole is covered with sagebrush steppe. Prime viewing spots include almost any pullout along Teton Park Road and Highway 191/26/89. Hikes along the valley floor almost invariably pass through sagebrush steppe, while those higher up along the mountains will give you views of this diverse ecosystem from above.**

STATE
MONTANA

ANNUAL VISITATION
3 MILLION

YEAR ESTABLISHED
1910

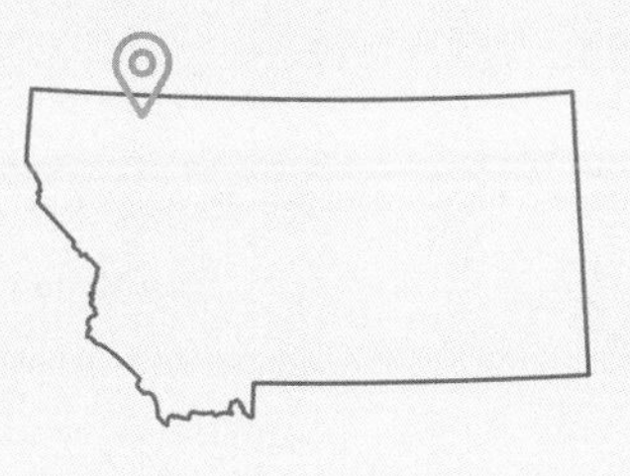

# glacier

## alpine vistas traversed by a thrilling road

| | |
|---|---|
| SUPERLATIVE | The stromatolite fossils found here are the oldest fossils in the national parks |
| CROWD-PLEASING HIKES | Forest and Fire Nature Trail (easy); Ptarmigan Falls (moderate); The Highline (strenuous) |
| NOTABLE ANIMALS | Mountain goat (*Oreamnos americanus*); common loon (*Gavia immer*); golden eagle (*Aquila chrysaetos*) |
| COMMON PLANTS | Western larch (*Larix occidentalis*); glacier lily (*Erythronium grandiflorum*); whitebark pine (*Pinus albicaulis*) |
| ICONIC EXPERIENCE | Snapping a photo of the mountains reflected in Lake McDonald |
| IT'S WORTH NOTING | In the 1930s, GLAC and Waterton Lakes National Park, just across the border in Alberta, combined to become the first International Peace Park in the world. |

**WHAT IS GLACIER?** GLAC is a magical, mountainous stretch of land in Montana. The park's signature roadway, Going-to-the-Sun Road, winds its way up the peaks and crosses the Continental Divide at Logan Pass. Get ready for tight curves, stomach-turning drops, and some of the best roadside scenery in North America. If you can peel your peepers off the peaks, spend some time exploring GLAC's beautiful lakes and streams. During busier months, consider riding the park shuttle instead of driving—it sure does make parking easier. Trip-plan fast enough and you might be able to hike to the last vestiges of the park's glaciers, predicted to melt away in the coming decades.

# GEOLOGY

## ANCIENT ALGAE

**WHEN LOOKING ACROSS GLAC TODAY,** you're probably awed by imposing mountains and sparkling lakes. But look closer and you'll come across fossils from a time when this place looked very different. One and a half billion years ago, this part of the world was covered in a tropical sea that stretched out from a barren rocky shore. The world was a harsh and different place then; there were no plants or animals on land and very little oxygen in the air. But in the shallower, sunnier parts of the sea, mats of tangled slime rose in reefs from the ocean floor. Known as *stromatolites*, these colonies of blue-green algae (also called *cyanobacteria*) were some of the first life on this planet. Even more crucially, stromatolites were the first creatures we know of to perform photosynthesis. Through that genius hack of turning the sun's energy into usable food, they pumped oxygen into the atmosphere and paved the way for the evolution of complex life. And it just so happens that GLAC is one of the best places on Earth to spot their fossils!

On the floor of that Precambrian sea, stromatolites started out as thin wisps of algae that shifting currents tangled up, the same way tiny threads on your sweaters eventually coalesce into little pills. Changing seasons and the algae's own chemistry trapped sediments between the layers, kind of like plaque between teeth. Those layers of sediment are what have been preserved as fossil evidence; only rarely do stromatolite fossils contain remnants of the algae themselves.

To spot the fossils, keep your eyes peeled for the swirls and eddies of those layers on rock outcrops and boulders. From the side, stromatolite fossils resemble the cross-section of a cabbage. Where they appear in groups, rock walls can suddenly begin to look like the sky from *Starry Night*. In many places you can stand next to a wall of ancient algae more than 10 feet high. In other places, you can walk over the top of stromatolite fossils; they look like closely packed, irregular mushroom caps.

---

You can see stromatolites actively growing today in the extreme environments of YELL's hot springs and geysers. Outside US national parks, living stromatolites are most famously present in Shark Bay, Australia, and in the Bahamas.

## SMALL, MIGHTY STROMATOLITES

Most stromatolite fossils look like wavy patterns in the rock. The easiest way to identify different types of stromatolites is by shape. Not all these characteristics are true for all specimens, but here are a few general things you can deduce from many GLAC stromatolites.

**CONICAL**

Slightly deep but still, calm water molded these stromatolites into conical shapes as they reached for the sunlight overhead. Conical stromatolites are uncommon today.

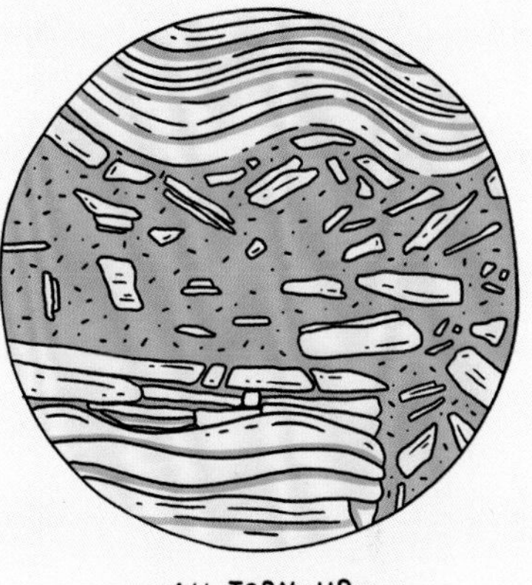

**ALL TORN UP**

Storms occasionally ripped apart stromatolite reefs and redeposited their torn-up bits elsewhere. Keep an eye out for fossils that look like they've been chopped up and whirled around.

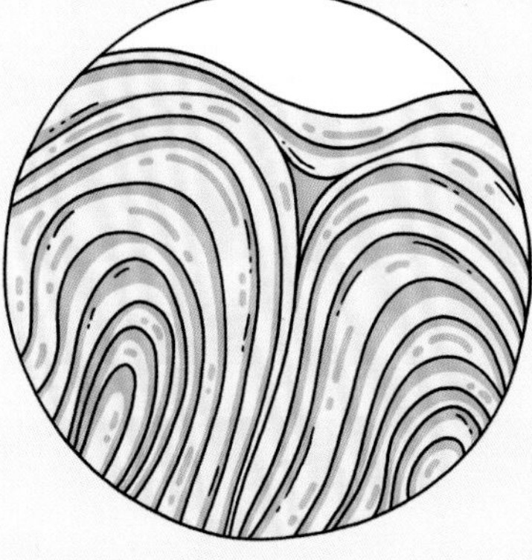

**MOUND**

The most common stromatolite fossil in GLAC. Quiet, calm waters let these stromatolites peacefully mound up. Some of their fossils are more than 5 feet across!

**COLUMNAR**

These stromatolites grew in the tidal zone. Their shape comes from being pushed around by currents and tides. Sometimes groupings of them can be used to study water direction and flow.

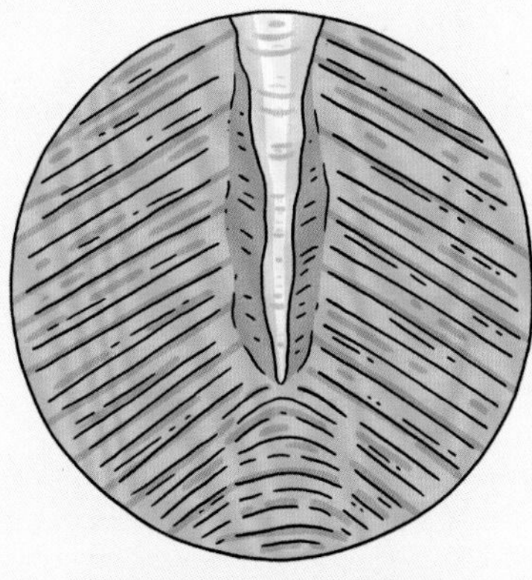

**BRANCHING COLUMNAR**

No one knows for sure why some columns branch. It could be related to changing water chemistry, current directions, an influx of sediments, a combination of all of these, or something totally different. To complicate matters, the majority of modern stromatolites are unbranched.

Stromatolites grew in a bunch of different shapes, depending on a host of factors such as water chemistry and depth. Look for mound-shaped fossils that resemble a cake with very thin layers. Some of them are more than 5 feet across! Many more are smaller—around a foot or so. Other stromatolites look like an opened Japanese fan. There are columnar stromatolites too—look for straight columns, as well as those with elaborate branches. One of the most striking groups of stromatolite fossils in GLAC are the conophyton. These cone-shaped stromies were so prolific that they have been preserved as a 100-foot-high band of fossils running through the park. These fossils don't erode as quickly as the rock that surrounds them, so look for sheer gray cliffs inside the heart of the park. Known as the Conophyton Zone, once you see it you'll be able to spot it from miles away.

Perhaps more important than their abundance was what the stromatolites gave off as a waste product. During photosynthesis, stromatolites took in the very abundant carbon dioxide on early Earth, used the carbon, and spewed oxygen out into the atmosphere. A friendlier, oxygenated atmosphere paved the way for the evolution of complex life, like you! So whenever you have one of those bad days that make you feel insignificant, remember the humble stromatolites. For them, small but mighty was enough to change the world.

**Roadcuts along Going-to-the-Sun Road boast great stromatolite fossils. Use turnouts and parking areas to get a better look—never stop on the road itself. Many trails feature fossils, including the Highline, Apikuni Falls, Grinnell Glacier, and Piegan Pass. See a ranger for more hiking suggestions tailored to your needs. Little Chief and Citadel Mountains and Matahpi Peak all host excellent exposures of the Conophyton Zone, which is also visible from the Lunch Creek parking area, about a mile east of Logan Pass.**

## WILDLIFE

### SAFETY AND PEE—A GOAT STORY

**NATIVE TO GLAC, MOUNTAIN GOATS** (*Oreamnos americanus*) are some of the most skilled mountaineers on the planet. Known for their sure-footedness, they are often seen hanging out on cliff edges, where no human would rationally be without ample safety gear and a racing heart. These white ungulates use their cloven hooves (split toes) to coolly navigate the cliff faces, which is especially helpful in the harsh mountain winters—they have access to flora that other animals can't get to, and they avoid predators at the same time.

As you're out hiking, look for mountain goats in the distance on the cliffs—they'll be white blobs, standing or resting on their bellies, likely with other goats. Get out those binoculars or cameras and have a nice long look. Keep your eyes open as you hike, though, because you might also see goats on the trails or even in parking areas, especially at Logan Pass and along Highline Trail; mountain goats have figured out the perks of hanging out with humans. Researchers dug into the Logan Pass goats' behavior after noticing that they're becoming more habituated than the more wild goats that live elsewhere in the park away from large numbers of humans.

There are two main benefits for the habituated goats. The first is where there are humans in great numbers, there are *no* grizzlies or other predators. It's more convenient than relying on steep cliffs for safety. The other benefit is salt. When humans pee along a trail, there's a salty residue after the pee dries that the goats lick up. Motor vehicle fluids also contain salt, and goats will lap those up from puddles in parking lots or roadsides. All goats need salt to stay alive, and female goats use salt in their production of milk. Goats in other parts of the park get it from natural sources, known as mineral licks.

Mountain goat (*Oreamnos americanus*)

Take a look around parking areas for trash on the ground—there might be goats nearby. Also, go pick up that trash.

Scientists were able to confirm how strongly the goats associated humans with safety when the Logan Pass area had to be closed due to a fire. Humans weren't around, although salt from their urine remained. Without humans, bear activity increased very quickly, and the goats retreated to the cliffs. Then, when the area reopened, humans returned, and with them came the goats, as bears retreated to less (human) populated areas.

When you're up at Logan Pass, look for a ranger named Gracie. She's got one brown eye and one blue eye, an orange safety vest, a collar, and the best attitude in town. She's a border collie, and she works to keep the goats (along with deer and bighorn sheep) wild by shepherding them away from heavily trafficked areas. Since she's reminiscent of a predator, she helps keep the ungulates away for a longer period of time. Sometimes the animals just need to see her and they'll avoid the area. As visitation increases, GLAC staff continues to work to manage human-wildlife interactions along with overcrowding (just like the other parks). Do your part, and make sure you stay far away from all of the wildlife in the park, including the goats. Also, remember to leave no trace and take all of your trash with you—that includes apple cores and banana peels!

**Anywhere there are mountain cliffs, there might be mountain goats. Keep an extra-sharp eye out on trails, like the Highline, and parking areas, like Logan Pass, because some goats are very comfortable with humans here. Report aggressive animal behavior right away to any park employee. Respect wildlife and keep a safe distance (at least 100 feet) from all wildlife, and 100 *yards* for bears and wolves.**

**HIKING IN GLAC IS HARD** work. It requires a lot of energy to roam mountain passageways, and yelling to warn bears of your presence can really dry out your throat. So when you aren't scanning the panoramic horizon, take some time to appreciate the scraggly understory shrubs growing alongside the trail. In the late summer and early fall, you may just be rewarded with a trailside treat that is practically synonymous with Montana: the huckleberry (*Vaccinium* spp.). Several species, all of which are edible, grow in GLAC. And it's perfectly within the rules to turn your fingers and tongue purple while snacking on them!

Huckleberries prefer forested areas with a relatively open canopy—look for forests of lodgepole, spruce, and fir. Like divas, huckleberries need quite a bit of light to thrive. Be on the lookout for steep mountainsides that funnel vegetation-clearing avalanches downhill and for areas that have been burned in a past forest fire. Hucks don't usually appear in the immediate aftermath of a blaze, but they go crazy in areas that were burned in the past.

These tiny bursts of juicy berry joy are a hell of a lot of fun to pick and eat trailside—go on, pop one into your mouth! But you aren't the only creature interested. Huckleberries account for almost 15 percent of grizzly bears' diets in the park. (Black bears love them too.) The berries' high sugar content really helps bears pack on fat before their hibernation. Keep an extremely sharp eye out for bears near huckleberries, and remember that *you* are in *their* home. If there are no bears around, look for their scat. In especially berry-rich years, autumn bear scat is often more huckleberry leaves than poo. Bears don't have the patience or the opposable thumbs to pick berries one at a time; they just go for the whole bush.

Several delicious species of huckleberries grow in GLAC.

## BOTANY

### HUMBLY YOURS, THE HUCKLEBERRY

Check in with a visitor center before sampling any wild wares and always carry a field guide! Black twinberries (*Lonicera involucrata*) look similar to huckleberries but grow in distinct pairs and are very unpalatable. They can be mildly toxic to some people.

The huckleberries here are related to both blueberries and cranberries—but so far these hucks have defied any effort to commercialize them. The only huckleberries you'll find for sale are wild-sourced.

Typically, berries ripen from late summer into early fall, but the effects of climate change are looming. Researchers in the park, aided by a battalion of citizen scientists (that you can join on any number of projects!), are working to understand how earlier springs, fluctuating temperatures, and changing precipitation will affect huckleberries and the species that rely on them. A mismatch between when the berries are ripe and when the bears need them will force bears to look elsewhere—especially in human-populated areas—for last-minute hibernation calories. The rhythm of the seasons and the timing of which plants and animals appear when is known as *phenology* to science—and as "we've been paying attention to that rhythm for millennia" to Native peoples. Understanding these cycles and how they are changing will be key in managing areas such as national parks in the future—and in many cases, citizen scientists will play a crucial role in data collection. Consider getting involved! GLAC's huckleberries are one key source of this climate-change information.

The humble huckleberry: stainer of fingers, tinter of tongues, bites for bears, and climate-change indicator par excellence.

**Huckleberries grow throughout the park. Each year the berry crop is different, so check in with rangers and local news sources about when and where the berries can be found. Park rules allow you to pick up to one quart of berries by hand per person per day. Waterton Lakes National Park in Canada allows only hand-to-mouth picking. Keep in mind that where there are huckleberries, there are also likely to be bears. Make plenty of noise, and keep children close by.**

**STATE**
**UTAH**

**ANNUAL VISITATION**
**2.7 MILLION**

**YEAR ESTABLISHED**
**1928**

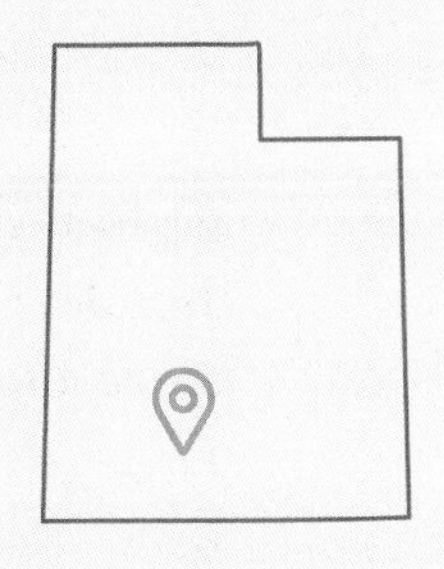

# bryce canyon

## high-elevation hoodoos & starry skies

| | |
|---|---|
| CROWD-PLEASING HIKES | The Queen's Garden Trail (easy); Swamp Canyon (moderate); Navajo Trail and Peek-A-Boo Loop (strenuous) |
| NOTABLE ANIMALS | Utah prairie dog (*Cynomys parvidens*); pronghorn (*Antilocapra americana*; see page 121); Great Basin rattlesnake (*Crotalus oreganus lutosus*) |
| COMMON PLANTS | Rocky Mountain juniper (*Juniperus scopulorum*); greenleaf manzanita (*Arctostaphylos patula*); Great Basin bristlecone pine (*Pinus longaeva*; see page 194) |
| ICONIC EXPERIENCE | Navigating the switchbacks of Wall Street |
| IT'S WORTH NOTING | The park offers ranger-led full-moon hikes through the hoodoos—the hikes have grown very popular, and space is limited, so plan ahead. |

**WHAT IS BRYCE CANYON?** It's not a canyon at all—it wasn't carved by a river, nor does it have two sides. Bryce "Canyon" is the rapidly eroding edge of a high plateau. Erosion has formed a series of natural amphitheaters here as well as the world's largest contingency of rock spires, called *hoodoos*. BRCA is the top step of the Grand Staircase, that immense and awesome series of colorful, horizontal rock layers that climbs upward from the Grand Canyon through Zion and into BRCA. The Grand Staircase is so big that you can't stand in any one place to take it all in, but here in BRCA, you can sit on the top step and ponder it all.

# BOTANY

## BERRY DECEITFUL

**TAKING THEIR CUE FROM BRYCE** (not a) Canyon, juniper trees are perhaps not the trees you thought they were. Their needles are scales, their berries are cones, and they are so much older than you think.

You've doubtless seen these trees or their close cousins in the park and in many other places as well. They live kind of everywhere in the Northern Hemisphere. Every family has its fun branch (pun intended); and for the junipers it's their European cousins, which have been used to flavor gin since the days of people donning metal suits and whacking each other with swords. The junipers that you see everywhere in BRCA have a much less wild, but also richly fascinating cultural history as a source of medicine, spiritual help, shelter, and food for the melting pot of Indigenous groups who live in this area.

There are three types of junipers in the park, but by far the most common is Rocky Mountain juniper (*Juniperus scopulorum*). Spot them at overlooks and alongside short trails into the amphitheater. If you're heading to either very high or very low elevations, you might also encounter common juniper (*Juniperus communis*) or Utah juniper (*Juniperus osteosperma*), respectively.

Take a moment to get up close and respectfully personal with a juniper tree. The needles are actually made up of a series of perfectly overlapping triangular scales that are just a few millimeters wide. Get out your loupe and have a look at those babies! Do you see the whitish lines between the scales? It's a waxy substance that helps

Like bristlecone pines (see page 194), junipers can self-prune and allow part of the tree to die, while remaining perfectly healthy elsewhere.

the plant retain more water. And it's a potent water-saving weapon when combined with those leaves-modified-into-scales and the juniper's crazily extensive roots. All that defensive strategy against water loss enables junipers to keep their leaves all year long. It's always photosynthesis time if you're a juniper.

And how about those berries! Look for bright bluish purple berries dusted in a waxy white coating. Those bursts of color are actually pea-size "pine cones" covered in a drought-resistant coating. Carefully break one apart to see some tiny but familiar-looking cone structures inside. Pop one into your mouth, but just one! It's like a concentrated burst of gin flavor without the hangover.

Juniper trees have such strong roots that the trees can continue growing even if they are knocked completely over!

Since you're standing by a tree with berries on it, you can be sure that the tree is at least 50, but probably closer to 200, years old; that's usually when they start producing seeds. Junipers typically live between 350 and 700 years, but can grow for a millennium or more. They grow slowly but with great determination and are often sculpted into craggy, unique shapes by wind, lightning, and shifting soil.

Spend a day exploring BRCA's junipers, then head back to your campsite and see if you can't rustle up a gin and tonic (you brought the ingredients, right?) to toast these ancient, berry-cool babes. You might have to forgo the ice, but these awesome junipers are worth a toast and so much more.

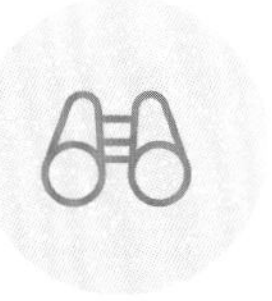

**You will likely encounter Rocky Mountain juniper at scenic overlooks and on short hikes into the amphitheater. Within the amphitheater, look for a mix of Utah and Rocky Mountain juniper. Common juniper is mostly found at higher elevations.**

# WILDCARD

## THE LONG VIEW

Notice that lights in the park are directed at the ground and shielded so that the light doesn't glare upward. You can make this happen in your community! Talk with your municipal leaders about "dark-sky friendly" lighting options.

**THE NIGHT SKY HAS ALWAYS** been important to humans. For most of our history it was our navigation system and our nighttime entertainment. But since the advent of electric light, after millions of years of dark night skies, most of us live where the artificial glare of street lights, businesses, advertisements, and our own porch lights takes over at sunset.

There are very few places left in the United States where you can still get a glimpse of our ancestral skies—but BRCA is one of them. Here, at high elevation and far from cities, you can see thousands of individual stars on any given night. Look through some binoculars or a small telescope and even more stars start to glitter and twinkle.

The show starts at sunset, when the colors on the hoodoos are as fantastic as you'd imagine. As the sky gets darker and darker, hundreds and then thousands of stars appear. Look up on clear nights during the summer and fall, and you'll get to see the Milky Way blazing across the sky. A hundred years ago it was not a terribly uncommon sight. Today, fully 80 percent of people in the United States live in places where light pollution drowns out the Milky Way. Take some time to appreciate that band of light. It's made up of billions of stars, maybe half of which host planets of their own. Who knows what (or who) is out there?

When you look at the Milky Way, you're looking at a tiny fraction of our home galaxy. We don't yet have the ability to send anything outside the galaxy to look back at us and send images, but we can see this one beautiful piece of our home. In a sense, the Milky Way is the only galactic mirror we have.

BRCA's phenomenal skies are brought to you by the same thing that also creates fantastic views during the day: namely, its location on a high desert plateau far from cities or industrial operations such as mines. The remote location cuts down on light pollution, and the thin, dry desert air doesn't transport or hold much in the way of particles. BRCA,

## SOAK UP THE STARS

BRCA has phenomenal skies, and the park works hard to preserve them. Here are some things you can do to maximize your time with the stars.

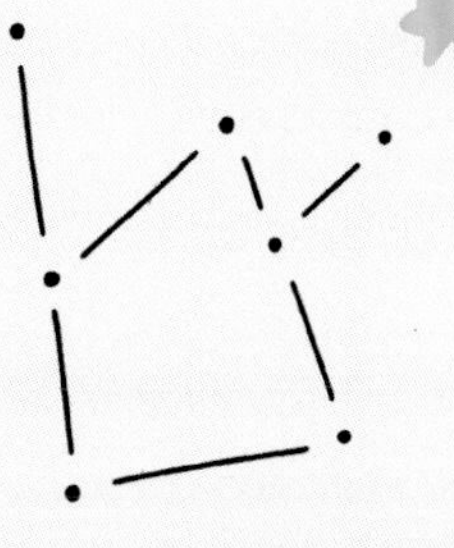

It takes about 20 minutes for your eyes to adjust to darkness. Be patient and you'll see so much more.

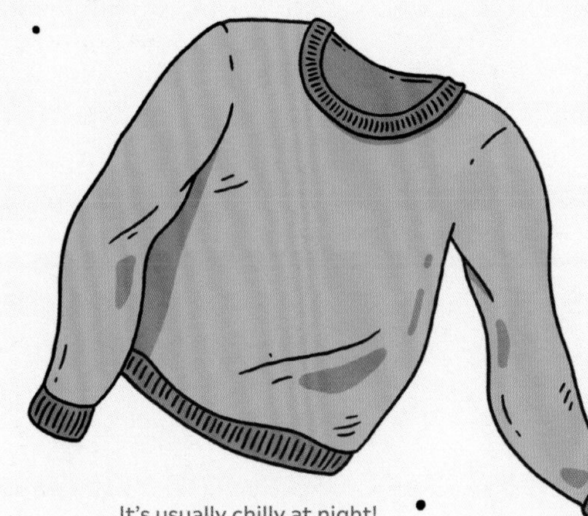

It's usually chilly at night! Dress in layers.

If you need light while outside, use a red light. In a pinch, secure red cellophane over your flashlight.

A decent pair of binoculars is as good as any small telescope—and usually easier to transport.

of course, cannot control what pollutants are blown in from outside, but the park has taken steps to reduce pollutants produced inside its boundaries—largely from automobiles. If you're here in the peak season, consider riding the shuttle—it will actually help your night-sky experience!

On a clear day, you can stand in BRCA and see almost 200 miles. At night, though, the field of view expands even farther away and even further back in time. Look for the Andromeda Galaxy. It's the most distant object you can see without magnification, but you should definitely grab some binoculars and take a closer look. What looks like a fuzzy, large star is actually a galaxy just slightly bigger than our own. It's the same shape too! It's so far away though—625 trillion times farther away from us than the moon—that you're viewing only the galaxy's super-condensed nucleus. The light that you're seeing left Andromeda 2.5 million years ago, just in time to join you here tonight. Don't waste it.

**BRCA hosts the longest-running astronomy festival in the NPS and was one of the first parks to offer organized night-sky programming. That tradition is going strong. Many volunteer organizations also set up telescopes for public use during peak season. (Pro tip: That happens in lots of other parks too!)**

## WILDLIFE

### A CURIOUS HUM

**THE SIGHT AND SOUND OF** hummingbirds ignite the imagination. Depending on where you live, you may have never had the chance to see (or hear!) these small creatures in real life. Hummingbirds are the second-most diverse family of birds in the Americas, and they're found only in the Western Hemisphere. While they're more abundant in Central and South America, more than a dozen species can be found in the United States. During your time in BRCA, you might be lucky to see one or more of the three common species that spend time in the park.

Hummingbirds are remarkable animals worthy of the curiosity they conjure. They're some of the best flyers on the planet, able to hover and fly backward. Their wings beat so fast (up to eighty flaps per second in some species) that they make the distinctive "hum" for which the birds are named. Most hummingbirds hover and use a long proboscis (tongue-like appendage) to drink nectar from flowers, and they often eat insects as well. Sometimes they even steal insects from spiderwebs.

Male broad-tailed hummingbird (*Selasphorus platycercus*)

The three species of hummingbirds commonly in the park are all around the same size, about 4 inches long with a 5-inch wingspan. The broad-tailed hummingbird (*Selasphorus platycercus*) is the one you'll see earliest in the spring, usually arriving in the park in April. Males are metallic green on their backs with red on their throats, and their hum is one of the most impressive of the family. It can be very loud and has a slightly metallic trill. Females are larger, also with metallic green on their backs and lines of green spots on their cheeks and throats. In the late spring, you might see the black-chinned hummingbird (*Archilochus alexandri*). Males and females are green; males have a velvety black throat with a thin, iridescent purple base, and females have a white chin. In the summer, rufous hummingbirds (*Selasphorus rufus*) appear. You can recognize them by the males' brilliant red-orange throat, rusty body, and white chest; females have an iridescent green back over a muted rust, with a white belly and some faint green spotting on the throat. Even if you can't identify which species you're seeing, you'll definitely know a hummingbird when you see it based on its aerial skills.

Rufous hummingbirds are the only hummingbird species known to migrate as far north as Alaska for the breeding season.

Like so many animals, hummingbirds may be key indicators of how our changing climate is affecting the landscape. There are many mysteries about these fantastic flyers, and work is being done to learn more about them, including their habitat preferences and their biology. Hummingbird banding and monitoring efforts have happened in BRCA and are under way in a lot of public lands of the West. As the years go by, more long-term information is logged. Since they're key pollinators, it is to our benefit to learn as much about them as possible and to ensure that we do our best to manage and maintain their habitat.

**Hummingbirds are often seen flying around the park meadows during the summer, looking for nectar plants.**

STATE
# UTAH

ANNUAL VISITATION
## 1.7 MILLION

YEAR ESTABLISHED
## 1971

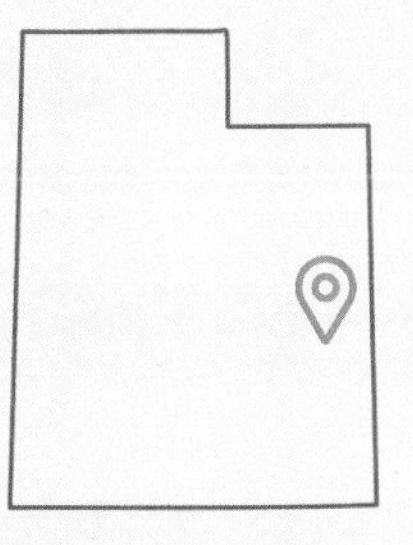



## iconic red rock wonderland

| | |
|---|---|
| SUPERLATIVE | ARCH has the densest concentration of natural stone arches in the world |
| CROWD-PLEASING HIKES | The Windows (easy); Park Avenue (moderate); Delicate Arch (strenuous) |
| NOTABLE ANIMALS | Raven (*Corvus corax*); mule deer (*Odocoileus hemionus*); western whiptail (*Aspidoscelis tigris*) |
| COMMON PLANTS | Utah juniper (*Juniperus osteosperma*; see page 134); sacred datura (*Datura wrightii*); dwarf mountain mahogany (*Cercocarpus intricatus*) |
| ICONIC EXPERIENCE | Seeing Delicate Arch |
| IT'S WORTH NOTING | You can join a ranger-led hike through the Fiery Furnace; check in at the visitor center to learn how. |
| SISTER PARK | Canyonlands (CANY) |

**WHAT IS ARCHES?** Stone arches, naturally carved by the elements over time, are the centerpiece of ARCH. Here in this sparse desert setting, the red rock seems to vibrate against the vast blue sky. The sandstone arches in the park range in size from sliver-thin cracks to spans more than 300 feet across. Be sure to check out the visitor center for a quick primer on how this intriguing scenery came to be. Spoiler alert: It involves a lot of salt! From there, the main park road traverses one amazing vista after another. Because wind, rain, and frost continue to sculpt this landscape, each visit will reveal new and unique sights.

# BOTANY

## IT'S ALIVE!

**DESERT AREAS ACROSS THE SOUTHWESTERN** United States don't seem like landscapes where life can thrive. Often, when we think of a place in nature flourishing, we consider lush forests, birds singing and flying overhead, mammals scurrying about, and bright colorful flowers. Desert life isn't necessarily as outwardly fertile or active, but it succeeds here too—it just has a different look.

The ground in ARCH, CANY, and many other desert areas on the Colorado Plateau (see margin note, page 100) is more than just dry sand. As you're walking in the park, look for dark, knobby, bumpy patches on top of lighter soil and sands. That is biological soil crust, also known as *cryptobiotic crust* or *desert glue*. It may not look like much, and sometimes you can't see it, but that crust is its own little world of micro activity.

Biological soil crust is found in arid environments all over the world, and in some places makes up about 70 percent of living ground cover. In this part of the country, it's made up of lichens, mosses, bacteria, and fungi. The star of the show is cyanobacteria, or blue-green algae (see page 126), some of the oldest known organisms on the planet. Cyanobacteria in desert environments build tiny filaments that activate when they get wet and move through the soil, creating a nice sticky sheath. Loose soil and rocks adhere to this "glue," creating the crust that we see. Over time, other organisms are able to take hold in this erosion-resistant surface.

The crust acts as a nitrogen- and carbon-fixer, providing a more nutrient-rich home for plants to grow. When it comes to erosion protection, it's pretty incredible too. Loose, dry soils blow away in the wind, ultimately creating dunes over just a few years. But cyanobacteria-dominated crust absorbs water more quickly and efficiently than loose soil, which reduces water runoff and holds the soil in place.

## A CLOSER LOOK AT CRUST

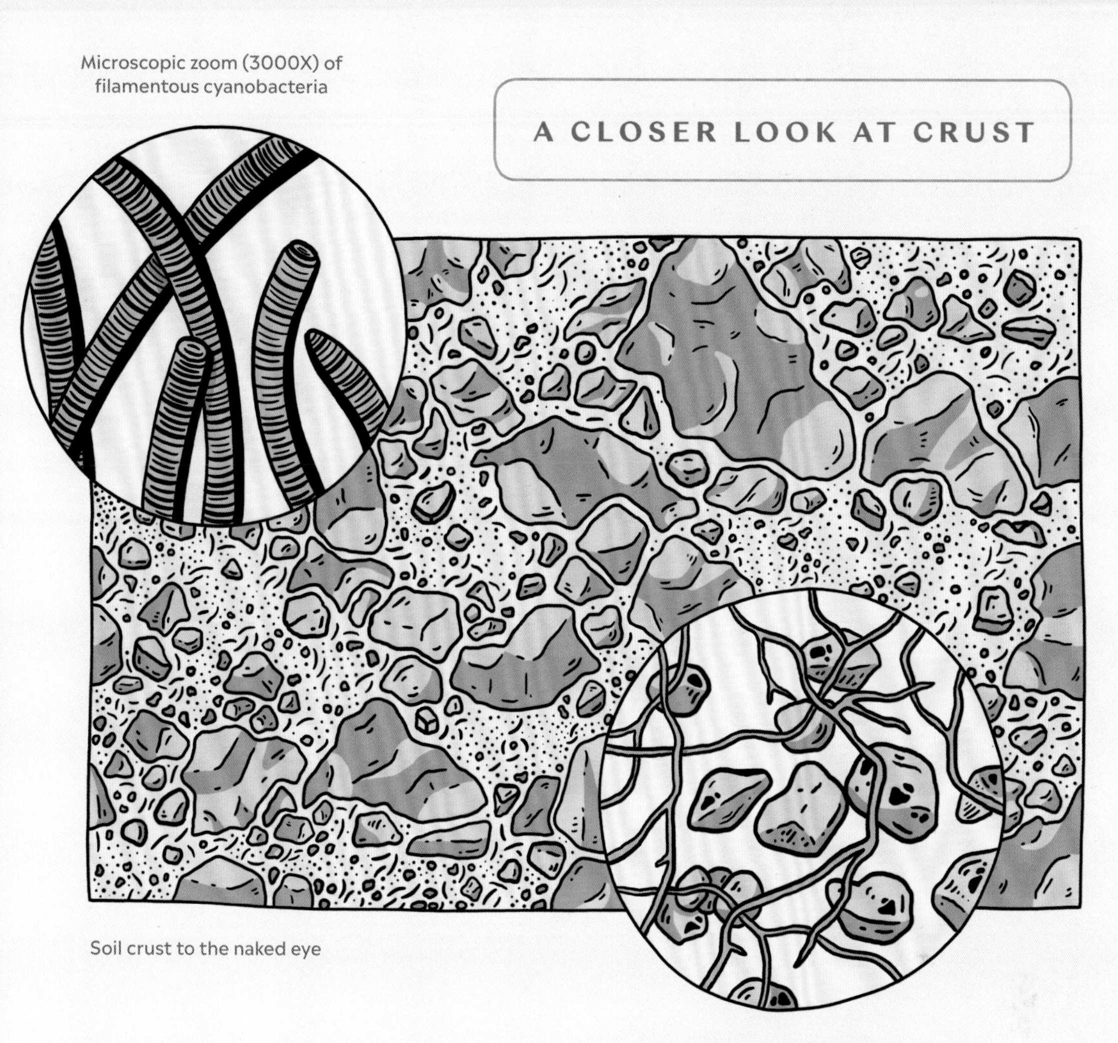

Microscopic zoom (3000X) of filamentous cyanobacteria

Soil crust to the naked eye

Filamentous cyanobacteria and bits of particles (magnified 90X)

These tiny worlds take years to form, and although they're mighty when it comes to protection from erosion, they're quite vulnerable to trampling by hiking boots, tires, and other human activity. It's crucial to stay on trails and stay off areas that have even the smallest amount of crust. In ideal circumstances, recovery from this sort of destruction can happen in several years. However, ideal circumstances aren't the norm, so it'll usually take decades—if not centuries—for a crust to form and even longer for lichen and mosses to take hold. Not only is this a problem for plants that use the crust to grow and the animals that eat those plants but it has huge implications for the land as a whole. If crust is destroyed and erosion is able to take full effect, much of this landscape has the potential to become moving sand dunes instead of a desert safe haven.

**Soil crusts are abundant in most desert parks—sometimes it'll be very easy to see, often dark and knobby looking. However, it's best to always stay on the trails to avoid disturbing delicate microscopic crusts.**

## WILDLIFE

### RAVEN GOOD-BYE TO A BAD REPUTATION

**WHEN YOU THINK ABOUT RAVENS** (*Corvus corax*), you might think about POEtry (get it?), acrobatic flying, unwanted pests, or who knows what else. Ravens are super-common in folklore, and they live around the world. They also happen to be some of the most intelligent birds on the planet, and their intelligence even rivals that of *human children* (more on that later).

Ravens are the largest songbird, about 2 feet tall, with a wingspan of a whopping 4 feet. They're larger than their counterpart, the crow, and have a bigger, curvier beak as well. There is a nice shininess to their black feathers, and they can be seen hopping or striding on land if they're not flying or hanging out at their large nests.

Raven (*Corvus corax*)

In the more heavily trafficked areas of ARCH and CANY (and other popular parks), you will likely notice how opportunistic ravens are when it comes to human food. They'll wait for people to leave their campsites, hoping for leftovers. You might see this as super-annoying, pest-like behavior. And, to be fair, it is, but it's also just one way you can see how intelligent these birds are. Of course, they'll grab any food that is sitting out in the open, but they'll also crack open coolers and unzip backpacks like it's their job. The real problem in the scenario is us—when we're better about keeping our food under wraps, ravens won't associate us with their next meal.

Research has been done on ravens raised in captivity, where scientists can give them scenarios that they'll never encounter in the wild (at least not that we can observe) to explore the limits of their intellect. Ravens are able to understand how to use a specific tool for a job to earn a food reward, something that most human children aren't able to do until the age of four years. They're also able to exercise restraint—such as waiting for a better prize instead of taking an immediate reward—also better than most human children. Often, when we think of clever creatures, we think of dolphins, primates, and other mammals, not birds. Ravens, we're sorry about that—we finally see you.

Remember to look up while you're out hiking, too, because ravens are just as impressive in the air as they are on the ground. You might see them playing with objects by dropping them and catching them before they hit the ground. If you see a couple of ravens in the air doing dips, rolls, and even flying upside down, it's possible that they're courting each other. They may even be a mated pair.

Ravens, like so many creatures in the national parks, have a lot more to them than a bad reputation as pests. As you safely lock up your food before your next hike, remember that the park is theirs as much as it is yours.

**Ravens are abundant in both ARCH and CANY, so you're likely to see them, especially in areas where humans hang out in large numbers.**

STATE
**UTAH**

ANNUAL VISITATION
**1.2 MILLION**

YEAR ESTABLISHED
**1971**

# capitol reef

## vibrant arches & high desert in red rock country

| | |
|---|---|
| SUPERLATIVE | The Waterpocket Fold is the longest exposed monocline in North America |
| CROWD-PLEASING HIKES | Capitol Gorge (easy); Hickman Bridge (moderate); Chimney Rock Loop (strenuous) |
| NOTABLE ANIMALS | Gopher snake (*Pituophis catenifer*); mule deer (*Odocoileus hemionus*); golden eagle (*Aquila chrysaetos*) |
| COMMON PLANTS | Prickly pear cactus (*Opuntia polyacantha*); Great Basin bristlecone pine (*Pinus longaeva*; see page 194); nakedstem sunray (*Enceliopsis nudicaulis*) |
| ICONIC EXPERIENCE | Cruising the Scenic Drive |
| IT'S WORTH NOTING | The park hosts an annual fruit harvest where the public are invited to pick their own cherries, apricots, and more. |

**WHAT IS CAPITOL REEF?** The "reef" of Capitol Reef refers to the fact that Euro-American settlers found this massive geologic feature to be a transportation barrier (like an ocean reef). In reality, it's a giant wrinkle in the earth's crust, where divots that erode in the rock fill with water. Sometimes called a *waterpocket fold*, these jumbled and uplifted rocks are the centerpiece of CARE. This is the heart of red rock country, complete with all the multihued rocks, arches, canyons, and cliffs you could ask for. When the sun goes down and the rocks go dark, look up to find a sky full of stars, planets, and the Milky Way.

# GEOLOGY

## RED ROCK COUNTRY DECODED

**UTAH MIGHT BE BEST KNOWN** for its stunning geology; you can't talk about Utah without talking about red rocks. This part of the United States is known as red rock country, but what makes the rocks red, anyway?

You might describe the vibrant red of the rocks as "rusty," which is spot on. Put very simply, there's a lot of iron in the rocks in CARE, and it has rusted. When iron is exposed to water and oxygen, it becomes iron oxide, which we call rust. There are several forms of iron oxide; the most common one in red rock country is called hematite. For the word nerds in the room, *hematite* comes from the ancient Greek word for "blood," *haîma*. (Hemoglobin and hemorrhage share the same root.) Tiny particles of hematite coat grains of sandstone, and it doesn't take much of the substance to make the rock radiate red.

Hematite isn't only found on Earth. Mars gets its nickname the "Red Planet" from the abundant fine grains of hematite in the dust that blows around during dust storms. Hematite doesn't only occur as tiny particles, though. There are darker areas on Mars, near the equator, where gray, more coarse-grained hematite is present. It isn't just about color—studying Martian hematite has produced a truly remarkable picture of that planet's ancient past.

Hematite on Earth forms in two ways: most often, as in CARE, it involves water and oxygen. It can also form without water, usually through the work of volcanoes. Mars has volcanoes, and it also has sedimentary rocks—but has it ever had water? And is that the origin of its hematite?

When Opportunity landed on Mars in 2004, the rover quickly began capturing stunning images of the Martian surface. One curious feature that kept cropping up in iron-rich areas were tiny spheres that appeared gray-blue in the images. The team started calling them "blueberries." If you spend any time in the more secluded parts of CARE, you just might come across some yourself! In red rock country they are known as *concretions*, and we know that they are a result of groundwater moving through iron-rich rocks, like the sandstone in CARE. The groundwater strips iron from the rock as it moves, stripping the rock of its rusty red color. When that iron-filled water runs into oxygen-rich water, the iron precipitates out into small spheres, discs, and cylinders. The same process is likely to have formed the blueberries on Mars—a crucial piece of evidence that the planet was once home to liquid water.

During your time in red rock country, feel free to let your imagination sweep you away to our celestial neighbor—there are many similarities between CARE and the Red Planet, after all. As research continues on Mars, take time to learn about it and enjoy the fact that the science happening on our world helps us understand not only our own geologic past but the past of our planetary neighbors as well.

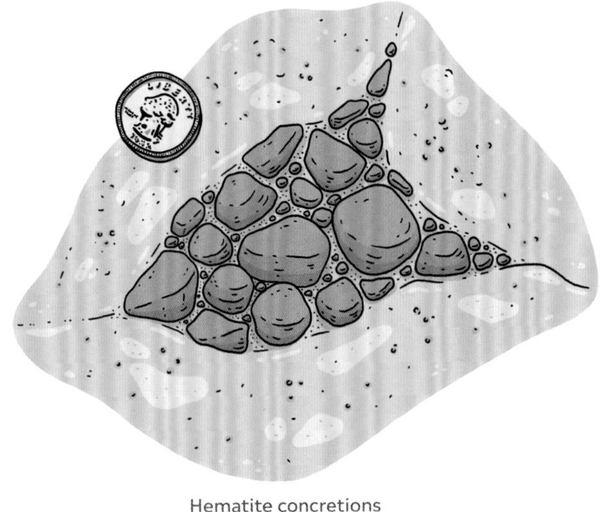

Hematite concretions
(spheres)

**Red rocks are abundant throughout the park. You can see hematite concretions in more secluded parts of CARE. If you spot any, be sure to leave them where you find them.**

# WILDLIFE

## RATTLING GOPHERS

**A VISIT TO RED ROCK** country should come with a plan to encounter a variety of wildlife, big and small. Snakes are just one kind of creature you might come across in CARE; conjuring up everything from fear to excitement to good old-fashioned curiosity in humans. The gopher snake (*Pituophis catenifer*) is common throughout western North America, and it's the snake you're most likely to see in the park. Fear not, though, because this serpent is harmless to humans, especially when you keep your distance and simply enjoy it with your eyes.

First, a little language lesson. *Poisonous* means "toxic when consumed"—that is, eaten; *venomous* means "toxic when injected by a bite or sting." In general, snakes are venomous, not poisonous. While gopher snakes can be confused with venomous rattlesnakes because of the sounds they create by vibrating their tails against the ground, their bites aren't venomous. Rattlers in this region are shorter and much stouter than gopher snakes. Rattlers also have larger heads, and black-and-white banding on their tails. These two species do look alike to the untrained eye, though, and when gopher snakes are threatened, they coil into a position similar to a rattlesnake's and look like they're going to strike. Instead of landing a venomous bite, they are known to strike with a closed mouth, which we'd like to call a *boop*.

Don't get close enough to any snake to find out whether you're going to get a bite or a boop.

As far as looks go, gopher snakes are pretty large. It's common to see adults reach 6 feet, and some can be as long as 8 feet! Their backs are cream or yellow with brown, black, or reddish blotches. Since they're not venomous, they don't kill their prey with their bite. Instead, they constrict their quarry and then consume it whole. Prey includes pocket gophers (obviously) and other small mammals, like moles and rabbits. Gopher snakes live in burrows and are accomplished climbers and swimmers.

## LOCOMOTION, NO LEGS REQUIRED

It can be hard to imagine how snake locomotion actually works. Use these illustrations to get a better understanding of how they use their bodies to navigate their world.

### LATERAL UNDULATION

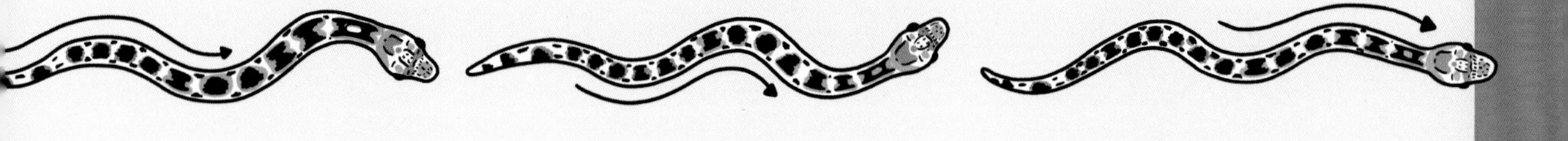

### RECTILINEAR MOVEMENT

### CONCERTINA LOCOMOTION

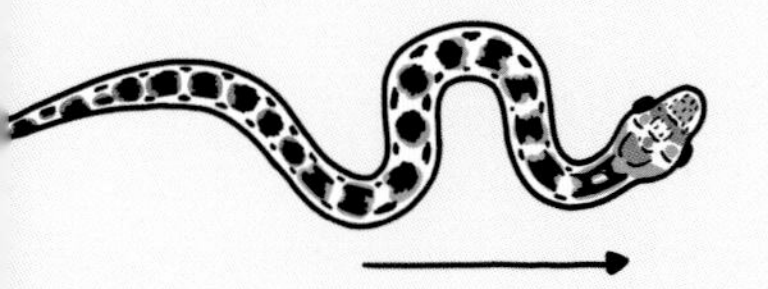

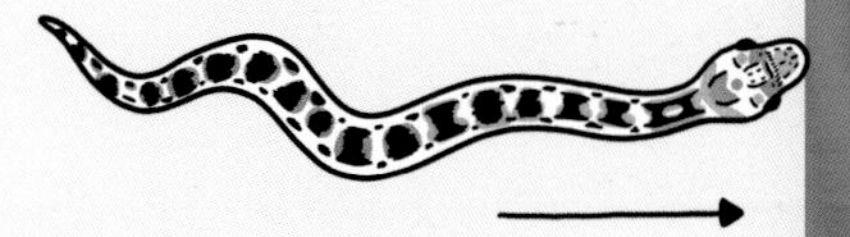

How do snakes navigate the world without legs, anyway? There are three common types of locomotion that gopher snakes use (although it's worth noting that there are other ways snakes move too). The first is the common *lateral undulation*, where the snake flexes its muscles to make a wave motion from head to tail, and its body pushes off of rough areas of the surface. They use a similar motion to swim. Second, gopher snakes move with *rectilinear movement*, which looks like they're creeping forward in a pretty straight line, kind of like a train. The snake uses muscles to scoot its belly scales forward, those scales connect with rough surfaces, and when the snake contracts other muscles, the rest of its body scoots forward. Finally, they have *concertina locomotion*, where the snake pulls its body into a series of bends, then straightens the bends to propel itself forward. Often, gopher snakes will use rectilinear or concertina locomotion to climb. As you're watching these snakes creep and slide their way through their day, remember that in the wild there's an adaptation for everything.

**Gopher snakes are common throughout CARE, and they're active during the day, especially at dawn and dusk. In the heat of summer, they're more active at night to take advantage of cooler temperatures. In the spring, males are more active in seeking a mate; and in fall, hatchlings emerge. Keep an eye out for these snakes, especially on roads, to make sure you don't injure or kill them.**

STATE
ARIZONA

ANNUAL VISITATION
957,000

YEAR ESTABLISHED
1994

# saguaro

## the classic cactus

| | |
|---|---|
| SUPERLATIVE | Saguaros are the largest cacti in the United States |
| CROWD-PLEASING HIKES | Loma Verde Loop (easy); Valley View Overlook (moderate); Hugh Norris Trail (strenuous) |
| NOTABLE ANIMALS | Greater roadrunner (*Geococcyx californianus*); regal horned lizard (*Phrynosoma solare*); javelina (*Pecari tajacu*) |
| COMMON PLANTS | Teddy bear cholla (*Cylindropuntia bigelovii*; see page 42); ocotillo (*Fouquieria splendens*; see page 182); creosote bush (*Larrea tridentata*; see page 53) |
| ICONIC EXPERIENCE | Marveling at the saguaros |
| IT'S WORTH NOTING | The eastern unit has older saguaros; younger stands grow in the west. |

**WHAT IS SAGUARO?** Saguaro cacti (*Carnegiea gigantea*) are an icon of North American deserts, though they grow only in a small area of *one* of those deserts. SAGU protects a stretch of their home range in two distinct areas that straddle Tucson, Arizona. To the west of the city is the smaller Tucson Mountain District, home to grasslands, shrubs, and plenty of wildlife. The Rincon Mountain District, to the east of the city, is larger and higher up. The increased elevation means sweeping views and conifer forests creeping down the slopes toward the desert below. Cruise the loop roads, take in the saguaros, and soak in the desert as you always imagined it.

# BOTANY

## GROWING HUMAN HISTORY

Many species of agave are hugely important to pollinators, like butterflies and bats with whom they have closely coevolved. Such is the case with the Palmer's century plant and bats of the genus *Leptonycteris*—which depend upon the agave as they migrate.

**THE SAGUAROS MAY BE THE** most well-known plant in SAGU's landscape, but it's worth looking around at their neighbors. There are some rather unassuming plants scattered throughout the park that are also expert desert dwellers and have shaped human cultures as much as humans have shaped them: the agave.

There are hundreds of species of agave, and at least as many hybrids and domesticated strains. Of those, about twenty are represented in the South and Southwest United States; three are easy to find and identify in SAGU. As succulents, agave plants have thick leaves that retain lots of water. They protect their precious stores with sharp teeth along leaf edges and other spiny adaptations. Often called *century plants*, most species of agave will flower and reproduce only once in their long lives. Their flowering stalks can grow to huge heights and look like giant, flower-topped asparagus spears—a handy way to remember that agave is actually in the asparagus family. Agave plants like stability. Unlike plants such as grasses, which help to stabilize whatever surface they grow on, agave won't move in until the ecosystem is settled and stable.

Being so full of water, agave plants cannot live where temperatures regularly dip below freezing—water expands as it freezes, destroying cell walls along the way. Ouch. Of the three kinds of agave you might spot in the park, Parry's century plant (*Agave parryi*) is the most cold tolerant. You'll find it growing in higher park elevations, where it serves as a rough indicator of the edge of the agave habitability zone. Look for thick, silvery green leaves that form a tight rosette growing close to the ground. Its strongly curved leaves give it the shape of a soccer ball.

Found at lower elevations, Schott's century plant (*Agave schottii*) is the most common agave in the park. You'll see it as low, untidy mats of thin leaves. Some patches get pretty large. Take a closer look and you'll see that the leaves taper

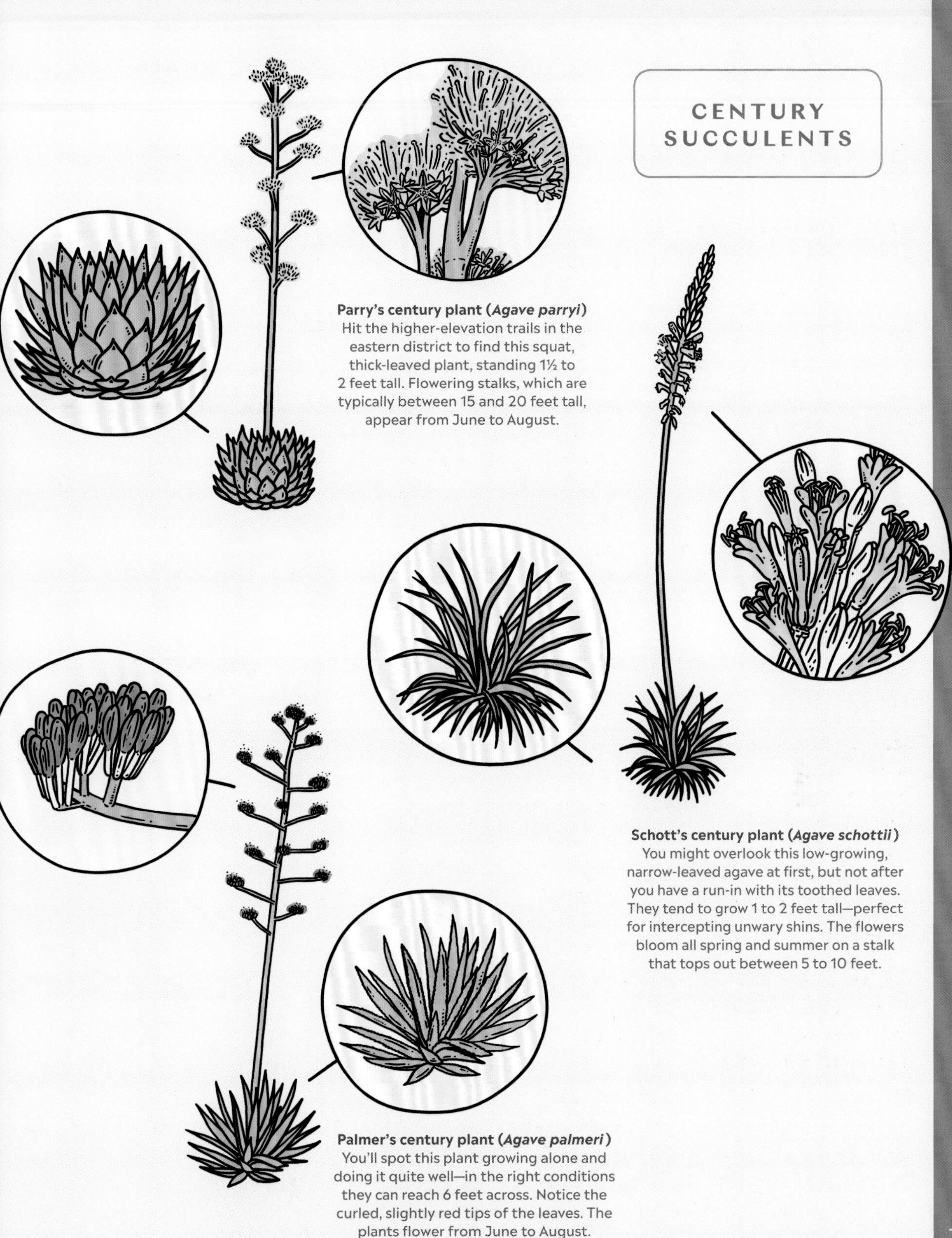

## CENTURY SUCCULENTS

**Parry's century plant (*Agave parryi*)**
Hit the higher-elevation trails in the eastern district to find this squat, thick-leaved plant, standing 1½ to 2 feet tall. Flowering stalks, which are typically between 15 and 20 feet tall, appear from June to August.

**Schott's century plant (*Agave schottii*)**
You might overlook this low-growing, narrow-leaved agave at first, but not after you have a run-in with its toothed leaves. They tend to grow 1 to 2 feet tall—perfect for intercepting unwary shins. The flowers bloom all spring and summer on a stalk that tops out between 5 to 10 feet.

**Palmer's century plant (*Agave palmeri*)**
You'll spot this plant growing alone and doing it quite well—in the right conditions they can reach 6 feet across. Notice the curled, slightly red tips of the leaves. The plants flower from June to August.

Today's O'odham Nation carries on desert traditions that have been going for thousands of years, like the harvest of saguaro fruit from plants inside the park.

toward the tip and are lined with fearsome teeth. This plant delivers unforgettable shin whacks to unlucky hikers. It's best to take time to appreciate them before you hit any hiking trails, as your curiosity may diminish once your legs have met this species.

The showiest of SAGU's agave is the Palmer's century plant (*Agave palmeri*). Growing alone in oak woodlands and rocky grasslands, this plant has narrow leaves that can reach almost 3 feet in length. If you can, get cautiously close and have a gander at the 2-inch spike that tips each leaf. The meaty centers of this species can be roasted and turned into alcoholic spirits such as tequila and mezcal. That's probably the best-known use of agave, but there are almost limitless others. Indigenous people of the Americas—including the O'odham, the original stewards of this land—have used agave for tens of thousands of years for food, medicine, shelter, and more. In the process of their long relationship, Native peoples created new strains and moved plants to new locations. It is actually impossible to separate the story of agave from the history of tens of thousands of years of human habitation in this part of the world.

Keep an eye out (and protect your shins from) these amazing plants as you explore SAGU. Agave offers a more understated sight than the tall, skyward-reaching forms of saguaros, but contained within its prickly profile is the deep knowledge of desert life and thousands of years of human-plant collaboration.

**Since SAGU is toward the northern limit of agave's range, lower elevations and sunnier slopes are good places to look. Schott's century plants are fairly widespread. Look for Palmer's and Parry's century plants as you hit the trails and gain in elevation.**

# WILDLIFE

## SMELL & THE GANG

**YOU'LL PROBABLY SMELL THEM BEFORE** you see them. They don't live in most of the United States, but are perfectly at home in the desert Southwest, including in both units of SAGU. Javelinas, as they are known in the United States and Mexico, are more properly called collared peccaries (*Pecari tajacu*), or stinky-cute-snorts if you're into nicknames that are not real or official.

Javelinas look a lot like pigs, but just as you wouldn't care to be mistaken for another kind of primate, javelinas would thank you to remember that they are actually not very closely related to pigs at all. The two animals started out together in Southeast Asia, but split off from each other between 40 million and 30 million years ago. Ancestors of modern javelinas got the travel bug and migrated across the Bering land bridge, becoming distinctly American animals. They were the only pig-like creatures here until Europeans introduced domesticated pigs during colonization.

Javelinas are smaller than pigs, have different teeth, lack tails, and have tusks that point down rather than out to the side. As with all wild animals, you'll want to give them plenty of space, so grab some binoculars to get a close look at their hind feet. Three delicate toes grace their back feet—a trait that puts them in the odd-toed ungulate category along

In South America, where the name *javelina* refers to a wild boar, these animals are called *peccaries*. This is a great example of why scientific names are useful and important (even though they aren't perfect).

Javelinas are sometimes struck and killed by cars in the park. Drive slowly and carefully to protect all wildlife (and yourself).

with rhinos, tapirs, and horses. (Pigs, with their four toes, hang out in the even-toed category with deer, sheep, and goats.) Take a gander at those highly mobile snouts as well. They look a lot like pig snouts and are used for many of the same tasks, like digging up roots and tubers, but are controlled by different bone and muscle structures.

You're unlikely to spot any lone javelinas. They are highly social and generally travel around in family groups, which range in size from a handful of individuals to gangs of twenty or more members. Regardless of group size, javelinas do team building really well. There is always a leader—usually a wise old female, because javelinas know what's up—and everybody wears a matching scent. All javelinas might smell the same to you—descriptions range from rotten cheese to skunk-funk—but to the javelinas themselves, scent is a clear and open way to communicate. Using special glands on their rumps (something else that distinguishes them from pigs), javelinas rub their scent on their surroundings and all over each other. Instead of T-shirts or bandanas, javelinas in a group coordinate their odor.

You won't see javelinas up and about during the day; like most desert dwellers, they are content to spend hot days lounging in the shade. Look for them to be more active early in the day or at dusk. They've been known to feed at night if it's really hot. If you're lucky, you may catch one snacking on a favorite javelina meal: prickly pear cactus, spikes and all. They might not be the pigs you were expecting, but they are amazing in their own right. Javelinas: big nose, big smell, big groups, big difference from pigs.

**Javelinas are fairly common in the park, including one group that regularly hangs out near Rincon Mountain Visitor Center. Look for them in the shade around dry washes.**

**STATE**
UTAH

**ANNUAL VISITATION**
739,000

**YEAR ESTABLISHED**
1964

## wild adventures among red rocks

**CROWD-PLEASING HIKES** Mesa Arch (easy); Aztec Butte (moderate); Syncline Loop (strenuous)

**NOTABLE ANIMALS** Turkey vulture (*Cathartes aura*); raven (*Corvus corax*; see page 145); bighorn sheep (*Ovis canadensis*)

**COMMON PLANTS** Claret cup cactus (*Echinocereus triglochidiatus*); blackbrush (*Coleogyne ramosissima*); pinyon pine (*Pinus edulis*)

**ICONIC EXPERIENCE** Driving (or biking! or motorcycling!) White Rim Road

**IT'S WORTH NOTING** Upheaval Dome's origin was once a complete mystery.

**SISTER PARK** Arches (ARCH)

**WHAT IS CANYONLANDS?** Made up of four districts, CANY is a sprawling, largely untamed maze of mesas, buttes, and canyons. Island in the Sky, an immense wedge of land squeezed between the Colorado and Green Rivers, is the most easily accessible and most visited district. From overlooks along its paved road, you can gaze down into a canyon 1,200 feet below. A thousand feet below that canyon's white rim are the rivers. Horseshoe Canyon, the Needles, and the Maze offer more seclusion and even more wild country. With few paved roads and even fewer services, CANY comes with huge payoffs for those prepared to rough it in red rock country.

## GEOLOGY

### DESERT VARNISH DREAMS

**IN DESERT LANDS THROUGHOUT THE** Four Corners region—including at ARCH and CANY—there are places where entire rock walls look like they've been masterfully painted with streaks of color. It's not paint, although it has been used by artists for thousands of years. The streaks you're seeing are desert varnish—a slowly accumulated, thin veneer of clay, minerals, and microbes on the rock. Its streaks and patterns are often something of a map for where water cascades over cliffs when it rains. You won't have to look very hard to find rocks decorated with desert varnish; in fact, it will be much harder to look away from these colors than to look for them.

Iron and manganese mixed with oxygen give desert varnish its colors. Red and orange streaks contain mostly iron minerals, while darker streaks contain more manganese. Keep an eye out for exceptionally smooth walls where manganese streaks take on a shiny, bluish tint. Streaks that are more brown or tan contain both iron and manganese. No matter the color, desert varnish contains particles of clay and vast numbers of microbes and fungi. If you pay close attention, you might notice that north- and east-facing walls have the most varnish, while south- and west-facing walls have far less. That distribution suggests that temperature has at least some influence on how and where desert varnish forms.

We still don't fully understand how it all comes together. Are the fungi and microbes that live on desert varnish actively helping to create it by oxidizing the iron and manganese, or are they bystanders? Answering that question has huge implications for the search for life on Mars, of all places. The Red

Desert varnish

Planet hosts some features that look very much like desert varnish. Could they contain some record of life there?

Desert varnish forms slowly, accumulating at a rate of less than the width of a human hair every thousand years. Rock walls change and erode over that course of time, so keep an eye out for spots where varnished rocks have broken away and left fresh rock exposed in seemingly random places.

Almost every Native culture that lived or still lives in North American deserts used desert varnish as a canvas for petroglyphs. With the help of the varnish's dark coating and lighter rock underneath, artists created works that still have the power to stop you in your tracks. Both ARCH and CANY contain petroglyphs carved into desert varnish. Should you be lucky enough to see some, you are welcome to look, respect, and wonder at these masterpieces. Never attempt to alter them or create your own.

Desert varnish adds one more layer of complexity and color to the rocks of ARCH and CANY. Its formation is a scientific mystery that might unlock extraterrestrial secrets. Over millennia, its slow, plodding growth has made it the perfect canvas for past rock carvings and present daydreams.

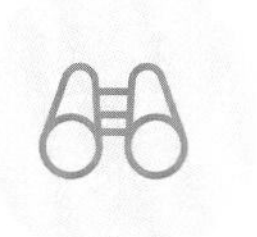

**Desert varnish is common on sandstone rock walls—those make up most of both ARCH and CANY!**

## DESERT POTHOLES: WAITING FOR A DRINK

AS YOU HIKE through CANY, keep an eye out for relatively level stretches of rock pockmarked with small basins. During dry times, you won't see much besides dirt and dry seedlings in these desert potholes. But when rain fills the depressions, they transform into miniature worlds with an incredible array of life: grasses, algae, worms, snails, tardigrades, and even tadpole shrimp that look like miniature horseshoe crabs. Be it wet or dry, never step in or disturb a pothole; your clunky feet can end the magic very quickly. Definitely take a good, long look, though. Here is one more example of life's endless tenacity and innovation.

STATE
**ARIZONA**

ANNUAL VISITATION
**645,000**

YEAR ESTABLISHED
**1962**

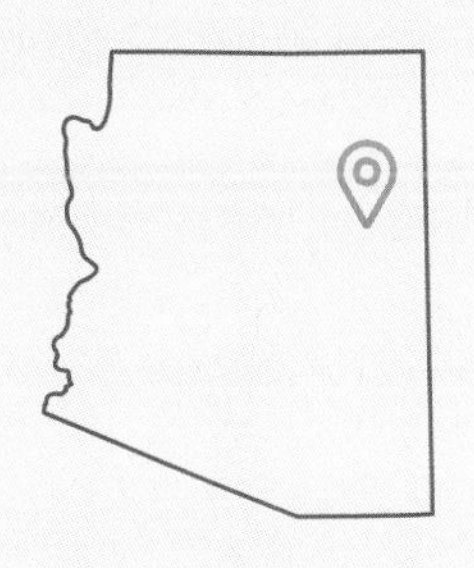

# petrified forest

## a geologic riot of color

| | |
|---|---|
| SUPERLATIVE | PEFO is the only national park site that contains a section of historic Route 66 |
| CROWD-PLEASING HIKES | Agate House and Long Logs Trail (easy); Blue Forest Trail (moderate); Onyx Bridge Trail (strenuous) |
| NOTABLE ANIMALS | Greater roadrunner (*Geococcyx californianus*); collared lizard (*Crotaphytus collaris*); pronghorn (*Antilocapra americana*; see page 121) |
| COMMON PLANTS | Needle-and-thread grass (*Hesperostipa comata*; see page 221); elegant sunburst lichen (*Xanthoria elegans*); banana yucca (*Yucca baccata*) |
| ICONIC EXPERIENCE | Getting your mind blown by the sheer quantity and bright colors of petrified wood |
| IT'S WORTH NOTING | At least a dozen types of plants that no longer grow in this area—including ginkgoes, conifers, and tree ferns—have been identified from their petrified remains here. |

**WHAT IS THE PETRIFIED FOREST?** Around 200 million years ago, this land was close to the equator and covered in a vast river system. Huge trees—some hundreds of feet tall—lined the banks. Trees that died and fell into the water were swiftly buried by mud and volcanic ash, eventually creating one of the largest concentrations of petrified (stonelike) wood in the world. PEFO is a multihued field trip through deep time. From the Painted Desert in the northern part of the park to brightly colored wildflowers and animal life, this is a world of endless colors and perpetual surprises.

## GEOLOGY

### PAINTING WITH ALL THE COLORS OF MINERALS

Silica naturally breaks cleanly along straight lines, which is why many of the logs look like they've been sawed into perfectly straight sections.

**THE PETRIFIED LOGS FROM WHICH** the park takes its name seem to be just about every color of the rainbow. Many of the logs are multihued, looking a bit like a stony tie-dye party. How does one tree get immortalized in so many different colors?

Wood becomes petrified when the element silicon dissolved in groundwater percolates through the tree's fibers, gradually filling in the empty spaces and replacing the once-living cells of the plant. For that to happen, the tree must be buried rapidly so it is cut off from outside oxygen and organisms that would normally consume it after death (yum). One of the most common forms of crystalline silica is quartz, which appears as clear or whitish chunks in the logs. Often though, silica contains very small amounts of other elements or minerals, which change the way light interacts with the crystals, and thus change the color you perceive. When a colorful petrified log catches your eye, it's those trace elements that call out to you.

Many of the most common colors—shades of pink, yellow, red, orange, and brown—reflect the plethora of ways that iron and oxygen interact. The most common is hematite (see page 148), aka rust. Microscopic inclusions of this mineral are responsible for the reds and pinks you see in the logs. The smallest pinch of carbon or manganese oxides produces darker shades of black and deep purple. Look closely, though, if the dark tint doesn't extend much beyond the surface, it might well be desert varnish (see page 160)—a thin veneer of minerals and microbes that builds up very slowly on hard, durable desert surfaces (like these logs).

What's so eye-catching about the petrified wood here is the kaleidoscopic colors—even a small piece of petrified wood can display a rainbow of hues. These different pigments are possible because wood doesn't become mineralized all at once; the process unfolds over a long stretch of time. Trees, just like you, are composed of many different kinds of

# THE PETRIFIED RAINBOW

Trace amounts of minerals in water and changing conditions over time have created a stunning number of mineral combinations—and bright colors—in the petrified logs.

**Chromium (Cr)** Uncommon, bright green; results from trace amounts of chromium. Darker green is usually the result of iron being *reduced* (mineralogist lingo for "deprived of oxygen").

**Manganese dioxide ($MnO_2$)** Purple and black; water traveling through volcanic rocks leaches manganese and redeposits it as a concentration of manganese dioxide inside the trees.

**Goethite ($FeO(OH)$)** Yellow, brown, and orange; this is another form of oxidized iron that is also called yellow or brown ochre for obvious reasons.

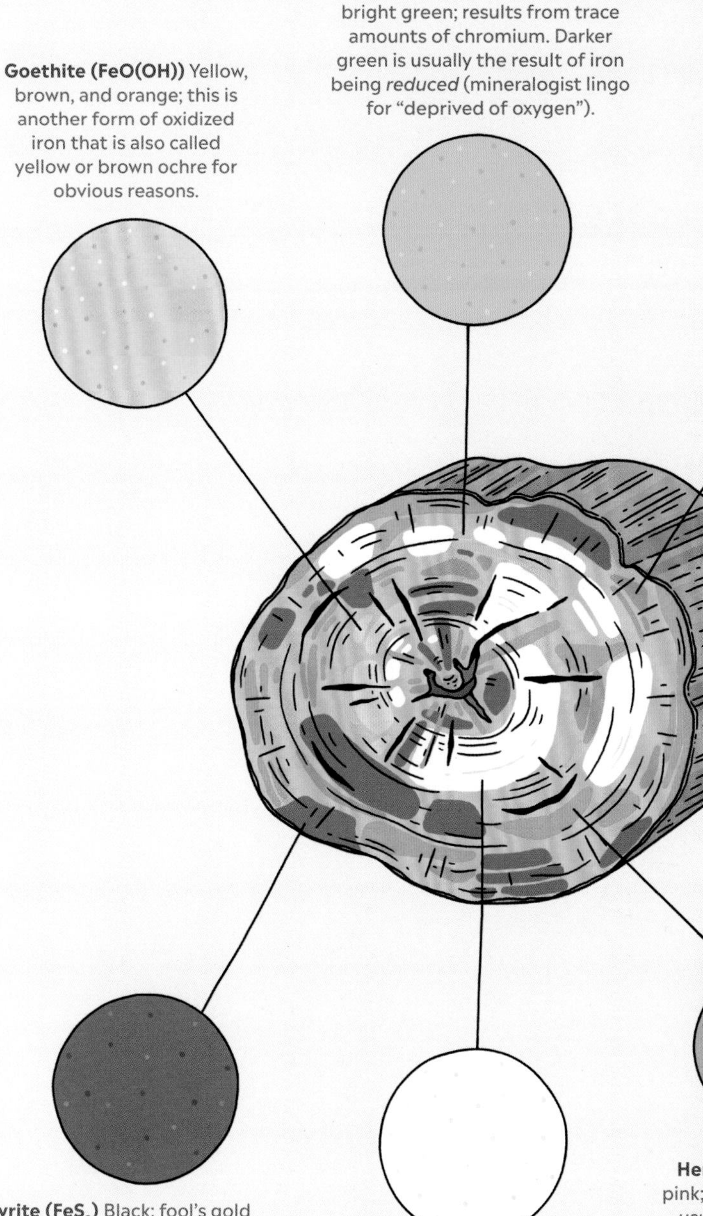

**Hematite ($Fe_2O_3$)** Red and pink; this kind of oxidized iron usually forms when iron in groundwater reacts (rusts) with the oxygen inside tree trunks.

**Pyrite ($FeS_2$)** Black; fool's gold forms in petrified wood when sulfur from decaying organic matter interacts with iron.

**Silicon dioxide ($SiO_2$)** White or clear; commonly found in the crystalline arrangement of quartz. Pure quartz is completely clear; the whitish color seen in some quartz is caused by many microscopic pockets of fluid.

The “rings” you might see on the logs are a result of gradual mineralization, not growth rings. The tropical climate of this area at the time when these trees were growing meant that they could likely grow year-round and thus wouldn’t have made annual growth rings.

tissues, and not all of them mineralize in the same way or at the same time. Over time, different elements make their way into groundwater and thus into the logs, where they react to form a colorful spectrum of minerals. What minerals form and when is also affected by changes in the temperature or pH of the water. Change the conditions, change the color.

Sometimes conditions during mineralization were just right to preserve extraordinary details within the wood. Some petrified wood in PEFO preserves the original cellular structure of the tree. It’s amazing to think that after more than 200 million years—a period of time longer than the existence of the Rocky Mountains and the Atlantic Ocean—it’s possible to see almost perfect copies of the cells of ancient trees.

All of the colors here are a link to not only that ancient period but also all the time between then and now. It’s a vast stretch that dwarfs human history and is a powerful reminder that even landscapes that look ancient and permanent are just glimpses of a much larger picture. Those bright streaks of color are the calling cards of deep time and ever-changing landscapes.

**Petrified wood is found throughout the park and is especially concentrated in the southern areas.**

# WILDLIFE

## NOT THE DINOSAURS YOU'RE EXPECTING

**THERE ARE DINOSAURS AMONG THE** petrified logs at PEFO. In fact, on many days you can see both living dinosaurs and dinosaur cousins hanging out, going about their business among the ancient trees.

First, let's talk about what a dinosaur is. Dinosaurs are a specialized kind of reptile, just as humans are a specialized kind of primate. Unlike every other reptile, dinosaurs have a special hole in their hip socket that allows their legs to be positioned directly under their body. Other reptiles hold their legs out to the side, in a kind of push-up position, forcing them to run with a side-to-side motion. Some can reach pretty respectable speeds, but none can sustain it for too long. Dinosaur hips let them move faster and farther, helping to ensure their long-lived success as a species.

By that definition, lizards—like the colorful collared lizards (*Crotaphytus collaris*) you might seeing sunning themselves in PEFO—are reptiles, but not dinosaurs. Grab some binoculars and get a good look at one: their legs are sprawled out to the side, with their thigh bones parallel to the ground. Collared lizards can really move when they need to; running on their back legs, they're capable of covering up to three times their body length with every stride, but only for short periods of time. Those long tails help them balance as they stride. The modern collared lizards you're seeing are virtually unchanged from 80-million-year-old fossils. That means that every single one of them is descended from ancestors who witnessed the catastrophic end of the non-avian dinosaurs and many other life forms as well.

Collared lizards are fierce predators with strong jaws who favor eating insects, but they're certainly not above snacking on smaller lizards.

The only dinosaurs who made it through the KT (or KPg) extinction event 65 million years ago were part of a specialized branch of feathered dinosaurs that would become modern birds. The common ravens (see page 145) that you definitely shouldn't be feeding at the picnic ground and the red-tailed hawks (see page 170) soaring above you *are* dinosaurs. The next time a bird is standing nearby, check out its legs; they are directly under its body, thanks to that special hip hole. Take a moment to count the toes; there are three of them, just like in the dinosaur footprints preserved all over the western United States.

Most visitors come to PEFO to see the Petrified Forest. But time exists in layers here, so all around the ancient logs are present-day animals. The lizards are modern creatures who still retain many of their ancestors' characteristics. The birds are only slightly more modern. They are the sole survivors among their family that once included the likes of *T. rex* and allosaurus—animals that would not look out of place here today.

**Birds and lizards are both fairly common sights in the park. Just keep your eyes open! Remember to always give animals space, and never harass or feed them.**

STATE
COLORADO

ANNUAL VISITATION
563,000

YEAR ESTABLISHED
1906

# mesa verde

## human history preserved in a stunning landscape

| | |
|---|---|
| CROWD-PLEASING HIKES | Farming Terrace Trail (easy); Petroglyph Point Trail (moderate); Prater Ridge Trail (strenuous) |
| NOTABLE ANIMALS | Broad-tailed hummingbird (*Selasphorus platycerus*; see page 139); mule deer (*Odocoileus hemionus*); collared lizard (*Crotaphytus collaris*; see page 167) |
| COMMON PLANTS | Douglas-fir (*Pseudotsuga menziesii*); big sagebrush (*Artemisia tridentata*; see page 123); showy daisy (*Erigeron speciosus*) |
| ICONIC EXPERIENCE | Taking a ranger-led tour of one of the Ancestral Puebloan cliff dwellings |
| IT'S WORTH NOTING | Mesa Verde is at a slight incline to the south, which helped the formation of the alcoves in which the cliff dwellings were built. |

**WHAT IS MESA VERDE?** Home to Ancestral Puebloan ruins, MEVE is an archaeological bastion on the Colorado Plateau (see margin note, page 100). This park was set aside to preserve this history, and it provides a rare glimpse into how the ancient Puebloan people lived. After you take a tour of the cliff dwellings, you can learn more about the present-day descendants of the Ancestral Puebloans in the visitor center exhibits. Head out for a hike to get a different view of the dwellings, and take in the plants and wildlife too. And be sure to hydrate—most hikes of any length feel strenuous here due to the altitude.

# WILDLIFE

## A SOARING SIGHT FOR YOUR EYES

**ONE OF THE MOST COMMON** and widespread raptors in North America spends time soaring overhead—the red-tailed hawk (*Buteo jamaicensis*). Odds are you've seen a red-tailed hawk at some point in your life if you've spent any time road tripping in the United States, Canada, or Mexico. These birds love to hang out on the tops of light posts near roads, looking for their next meal with amazingly acute vision. Look up during your time in the park—you'll probably see the occasional bird in sky, and if it's flaplessly riding the thermals (a column of air rising in the atmosphere), it's quite possibly a red-tailed hawk.

Obviously, if you see a red or rust-colored tail, that's a telltale sign that you're looking at this common raptor. If the tail color is hard to determine, look at its belly—red-tails usually have a pale belly with a dark breast band. Their wings typically have a dark bar on the leading edge from shoulder to tip. However, color can really vary, and there are some red-tails with a reddish or chocolate-brown belly. Look at the shape of the bird's body—if it's our friend the red-tailed hawk, it'll have a short tail and broad, rounded wings. Look at how it flies—if it's flapping a bunch, it's a different bird. If it's soaring peacefully, like it has nowhere to be, it's probably a red-tailed hawk.

Red-tailed hawk (*Buteo jamaicensis*); perched

During the chick-rearing season in early spring, males and females build their nests together, and typically the males will bring back food for the female and their babies. Red-tailed hawks don't necessarily stick with the same mate, but they are likely to stick to a particular nesting site. Often, their mate also returns, and they stay together and use the same site year after year. They're opportunistic hunters, typically eating small rodents, and aren't above eating carrion (already dead animals, like roadkill). To mix it up, they'll also eat lizards and reptiles on occasion, and some intrepid hawks have even been known to kill a rattlesnake.

Red-tailed hawk (*Buteo jamaicensis*); in flight

As you're meandering through the park (and anywhere in North America, really), keep your ears open too. *What am I listening for*, you ask? Close your eyes and imagine the sound of any hawk or bird of prey you've seen in a movie or TV show. That typical hawk-like cry you're imagining is more than likely the sound of a red-tailed hawk. Years ago, Hollywood decided that its sound was the most predatory bird-sounding of them all, so now sound designers toss in that red-tailed hawk call during postproduction and call it a blockbuster. Tell all your friends in the park about it on your next hike, and use your reliable eyes and ears to find some fabulous birds!

**The easiest way to see a red-tailed hawk is to look up at possible perches that offer high vantage points, where they'll be scanning the ground for food. They also may be riding thermals throughout the park.**

and another thing

## ARCHAEOASTRONOMY: CONNECTING ANCIENT PEOPLE WITH THE NIGHT SKY

***ARCHAEOASTRONOMY* IS THE** study of how past peoples interacted with the sky, and it's one discipline that's been used to interpret the ways the Ancestral Puebloan people may have lived. In MEVE, two sites in particular have apparent connections to celestial bodies: Cliff Palace and Sun Temple. Sun Temple, for instance, is aligned to the winter solstice sunset and also appears to be connected with the moon and Venus. When you look up at the sky, remember that humans have been curious about our universe for as long as we've walked the planet.

# carlsbad caverns

**STATE**
NEW MEXICO

**ANNUAL VISITATION**
466,000

**YEAR ESTABLISHED**
1930

## elaborate caves hidden under a desert surface

**SUPERLATIVE** Lechuguilla is the deepest cave in the United States

**CROWD-PLEASING HIKES** Chihuahuan Desert Nature Trail (easy); Rattlesnake Canyon (moderate); Slaughter Canyon (strenuous)

**NOTABLE ANIMALS** Brazilian free-tailed bat (*Tadarida brasiliensis*); Couch's spadefoot (*Scaphiopus couchii*); cave swallow (*Petrochelidon fulva*)

**COMMON PLANTS** Torrey yucca (*Yucca torrey*); cloak fern (*Notholaena* spp.); curlyleaf muhly (*Muhlenbergia setifolia*)

**ICONIC EXPERIENCE** Watching the nightly bat flight from the outdoor amphitheater at the natural entrance to Carlsbad Cavern

**IT'S WORTH NOTING** You can explore two parts of Carlsbad Caverns at your own pace using Big Room Trail and Natural Entrance Trail.

**WHAT ARE CARLSBAD CAVERNS?** Laid down more than 250 million years ago with the same reef formation as the Guadalupe Mountains, CAVE is a massive series of more than a hundred limestone caves located underneath the Chihuahuan Desert. Filled with cave creatures such as crickets and bats, Carlsbad Cavern itself is the main attraction. Its distinctive limestone features were carved by acid, not water, and are filled with thrilling gypsum deposits that aren't easily forgotten. Aboveground, you can see abundant wildlife year-round, including plenty of resident and migratory birds. Bring your binoculars and your loupe and have a long look around this magical desert landscape.

# WILDLIFE

## BIRDS WITH A CURFEW

**ONE OF THE MOST SPECTACULAR** experiences in CAVE is watching the hundreds of thousands of bats (more than 400,000 roost in Carlsbad Cavern) emerge every evening at dusk from late spring through autumn, with the best viewing in the summer. If you're willing to get there during the day, before sunset and the bat flight, you can enjoy an airborne avian show. Cave swallows (*Petrochelidon fulva*) breed at the entrance to Carlsbad Cavern, and seeing them is worth the visit.

Cave swallows are the less common cousins of cliff and barn swallows, and, in the United States, most are found in New Mexico, Texas, and Florida. Fortunately for visitors to CAVE, a large colony breeds year after year in Carlsbad Cavern before heading south for the winter. Cave swallows typically nest in caves (hence the name) or sinkholes, but they have settled under bridges or other human-made structures as well. The cave swallows here use mud and bat guano to build their open, cup-shaped nests, attached to cave walls. These can be used year after year, as long as they're protected from the elements. Don't worry, the swallows line the nests with feathers and other fluff, so those babies are comfortable.

When you visit the swallows' breeding area, listen to their calls and songs. Typically, their calls are short whistles that sound like "zreet zreet" and are far less complicated than their songs. Their songs are believed to relate to courtship, and include squeaks and warbles along with knocks and gurgles. While it may not sound like a beautiful ballad to us, it gets the job done in this colony, which continues to breed here every year. In terms of flight, you'll see swallows darting in and out of the cave all day long, looking for a delicious insect feast.

Cave swallow
(*Petrochelidon fulva*)

Interestingly, cave swallows prefer to eat in flight, as opposed to stashing food for later. As the sun sets, the cave swallows go silent and head to the cave for the night, giving the limelight to the bats.

If a cave swallow doesn't get home in time, it can take hours before the stream of bats dissipates enough so they can get back in. Talk about a strict curfew!

Cave swallows don't live in CAVE year-round, and it was a hefty undertaking to solve the mystery of where they winter. In the 1980s, an intrepid researcher named Steve West gained permission to band the birds in the park to try to sort this out. He and many other volunteers banded birds every year, in the hopes that they'd find them wintering in the caves in Mexico. Years passed, and they gathered as much data as possible on the birds as they banded them, but they never found the birds in Mexico. Persistence won, though; in 1992, a banded swallow turned up (dead, unfortunately) in western Mexico, which led researchers to look for the birds farther west and south. In 1994, they were found in El Salvador, living *not* in caves but among sugarcane, mangroves, and grain crops.

Even though the wintering mystery has been solved, banding continues every year, with the help of volunteers, and more questions are answered. Since the banding has gone on for decades, there is data about the swallows' lifestyles, lifespans, and more. The cave swallows are just one example of research persistence being rewarded, underscoring that we never know everything about an animal species. Sometimes, the more we know, the more we realize we don't know.

The large colony of cave swallows in the park can be seen during warmer months flying in and out of the Carlsbad Cavern entrance throughout the day. Ask at the visitor center about what research is happening currently with the birds, and check whether there are volunteer opportunities to join in on the banding!

## and another thing

## THE CAVE OF CAVES: LECHUGUILLA

**CAVE IS HOME** to more than a hundred caves, but one takes the cave cake as the most remarkable, and it might even be the most stunning cave in the whole wide world. Lechuguilla's massive size (149 miles long and 1,604 feet deep, the second-deepest cave in the United States) has allowed for colossal gypsum and sulfur deposits to form in an untouched environment. Since it's a rare geologic treasure, the only people who can visit the cave are researchers with permits (rest assured, they've taken photographs for you to see). After visiting Carlsbad Cavern, though, it's not hard to imagine what else is waiting to be discovered in this geologic wonderland.

**STATE**
COLORADO

**ANNUAL VISITATION**
443,000

**YEAR ESTABLISHED**
2004

# great sand dunes

## giant dunes & beautiful mountains

| | |
|---|---|
| **SUPERLATIVE** | GRSA is home to the tallest dunes in North America |
| **CROWD-PLEASING HIKES** | Sand Sheet Loop Trail (easy); High Dune (moderate); Star Dune (strenuous) |
| **NOTABLE ANIMALS** | Western tanager (*Piranga ludoviciana*); pronghorn (*Antilocapra americana*; see page 121); circus beetle (*Eleodes hirtipennis*) |
| **COMMON PLANTS** | Scurfpea (*Psoralidium lanceolatum*); quaking aspen (*Populus tremuloides*); Rocky Mountain beeplant (*Cleome serrulata*) |
| **ICONIC EXPERIENCE** | Running, boarding, sledding, or jumping down the dunes |
| **IT'S WORTH NOTING** | Mount Blanca is more properly called *Sisnaajini*, one of four sacred peaks that bound the homeland of the Diné (Navajo). |

**WHAT ARE THE GREAT SAND DUNES?** The dunes here are the dried-up remains of an ancient lake, blown up against a beautiful and rather tall mountain range, the Sangre de Cristo. GRSA includes a national park that protects the dunes themselves, as well as the adjoining grassland that grows from a giant sheet of sand. The adjacent national preserve protects the mountains—an important part of the complex dune environment. The dunes themselves offer ample room to gleefully explore a surface that is continually rewritten. Hike the dunes and splash in the seasonal mountain stream, but don't neglect the grasslands and mountains; there is much more to GRSA than meets the sand.

# BOTANY

## SANDY SUNFLOWERS

**DURING LATE SUMMER, SUNFLOWERS CARPET** the grasslands and spring up from the dunes in GRSA. The prairie sunflower (*Helianthus petiolaris*) is native to this area, and indeed to much of the western United States. If you drive to Colorado from the east, you can watch common sunflowers (producers of the snack seeds) along the road transition into prairie sunflowers in the western plains. North America is the original home of sunflowers, with a number of species spread around the continent, but only the prairie sunflower is able to make its home in the sandy, shifting world of the dunes.

Prairie sunflowers, like all sunflowers, are annual plants, which means that they grow, flower, and produce all the seeds they ever will within one growing season. That could be a very risky strategy; if all the seeds attempt to germinate and some natural disruption happens—say, a late frost or below-average rainfall—entire populations of sunflowers would be wiped out. These flowers, though, have come up with an ingenious strategy. A small percentage of their seeds will germinate no matter what, and a much larger percentage (nearly half in some cases) will germinate given good conditions in the spring. A final small percentage seem programmed to remain dormant as backup stock. Were something, like extreme weather or disease, to wipe out every standing flower, more would still be able to grow the following season. Seeds can remain dormant for a decade or more, though what exactly triggers some seeds to germinate while others bide their time remains a mystery.

Sunflowers are probably best known for tracking the sun, turning to face it throughout the day. Actually, only immature flowers track the sun—mature blooms face east all dang day. This sun-greeting position allows them to warm up quickly in the morning, making them attractive to pollinators, while avoiding the harshest temperatures of the midday sun.

Take a moment to check out the plant's hairy leaves and stems. They help reduce water loss in this dry, high-elevation environment.

## SUN ON THE PRAIRIE

Prairie sunflowers, like all sunflowers, are composite flowers; their blooms are actually composed of many tiny flowers all clustered together into one larger bloom.

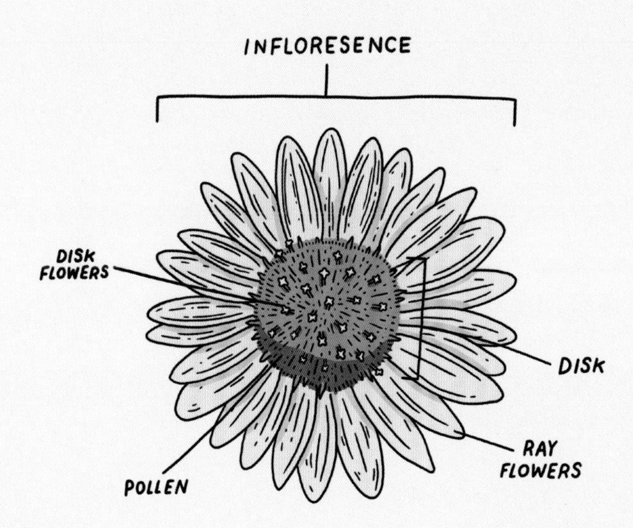


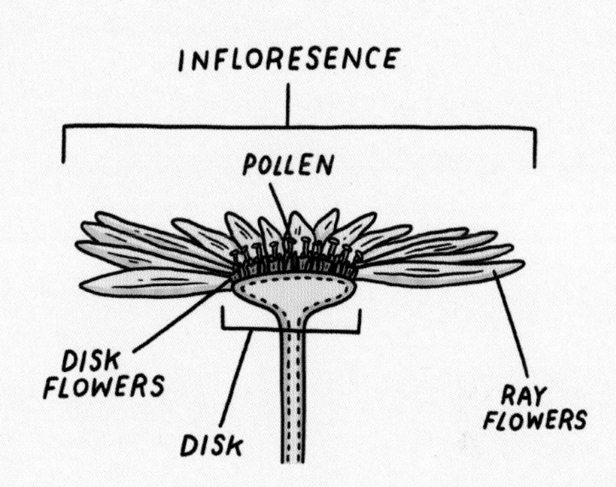


PRAIRIE SUNFLOWER
(*HELIANTHUS PETIOLARIS*)

In exceptionally wet years, it can seem like there are uncountable numbers of sunflowers in the grass and sand in GRSA. But there are even more flowers than you know! Take a moment to get up close with a sunflower. They are not one flower, but are more properly called *inflorescences*—many smaller flowers, known as florets, clustered together into seemingly one big flower. Prairie sunflowers have two types of florets that you will quickly be able to tell apart. Ray florets develop into the yellow "petals" you associate with these flowers. Look closely at the center ring of the bloom,

Recent studies have shown that sunflowers that grow on the dunes are developing traits different from their grasslands counterparts. It's possible that they are becoming a distinct species!

it is filled with much smaller disc florets. These tiny, tube-shaped flowers are what produce the plant's seeds. See any yellow dots? Those are balls of pollen held aloft for the bees, flies, moths, and other winged critters that help pollinate these plants.

The stark contrast of immense dunes backed by mountains is part of what draws visitors to GRSA every year. It's a wonder that any plants manage to survive in the sand. Even a short hike can make your calves scream and your level of appreciation for dune plants grow exponentially. Sunflowers add a beautiful splash of color to the dunes and their surrounding grasslands. But now you know that taking these flowers at face value ignores the multitudes they contain and the meticulous strategy that goes into their yearly showcase of color and pollen.

**You can see fields of prairie sunflowers in the grasslands and more isolated blooms in the dunes from June to September.**

STATE
TEXAS

ANNUAL VISITATION
440,000

YEAR ESTABLISHED
1944

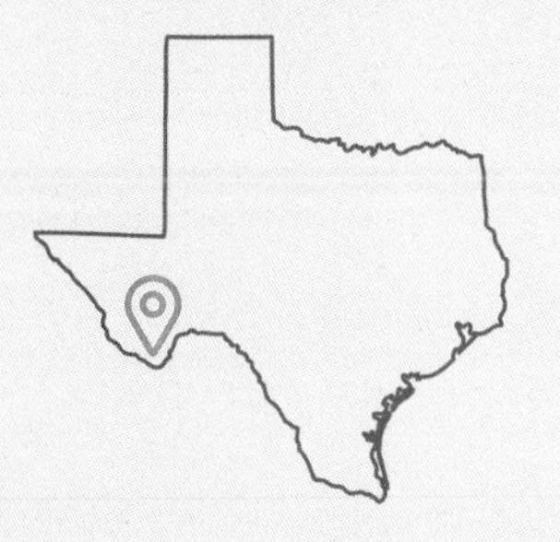

## the mountains at the end of a desert road

| | |
|---|---|
| SUPERLATIVE | BIBE is home to more species of cactus than any other national park |
| CROWD-PLEASING HIKES | Hot Springs Historic Trail (easy); Mule Ears Spring (moderate); Emory Peak (strenuous) |
| NOTABLE ANIMALS | Blue-throated hummingbird (*Lampornis clemenciae*); Merriam's kangaroo rat (*Dipodomys merriami*; see page 47); javelina (*Pecari tajacu*; see page 157) |
| COMMON PLANTS | Harvard's agave (*Agave havardiana*); ocotillo (*Fouquieria splendens*); honey mesquite (*Prosopis glandulosa*) |
| ICONIC EXPERIENCE | Paddling or walking into Santa Elena Canyon on the Rio Grande |
| IT'S WORTH NOTING | This remote desert park boasts some of the best night skies in the country. |

**WHAT IS BIG BEND?** The "big bend" is a 90-degree turn in the course of the Rio Grande, which makes up the southern border of both BIBE and the United States. You can't look out at the bend itself from the park, but rest assured there is plenty else to explore. The park protects not just river habitat but also stretches of the Chihuahuan Desert out of which the Chisos Mountains rise. Their cooler environs are like a biological island in the sky. Five visitor centers serve this sprawling park, where one of North America's defining rivers carves through sandstone in a landscape of parched desert and verdant mountains.

# BOTANY

## DESERT TORCHES

**THROUGHOUT THE DESERT OF BIBE,** and much of the southwestern United States, an eye-catching plant stretches spiny, spindly, usually bare branches up toward the unforgiving sky. Ocotillos (*Fouquieria splendens*) often look dead, but these desert shrubs are experts at biding their time and conserving their energy. When the moment is right, seemingly dead sticks transform into columns of green tipped with fire.

Stop and examine an ocotillo near a visitor center or next to a hiking trail. Most of the year, they resemble a strange, chaotic jumble of dead sticks. A dozen or so sharp-spined, white-and-gray stems radiate out haphazardly from the ground. Looking closely will reveal dark furrows and possibly even a hint of green along the stems. These spires can grow up to 20 feet tall—they look a little out of place among the mostly low-growing plants of the desert. The stems store water and are capable of photosynthesis, and thus the plant protects them with spines. Take a close look at a spine; it is actually the leftover base from the ocotillo's first set of leaves. Nestled near the base of the spine are structures that will produce every set of leaves the plant will ever grow.

Whenever there is enough rain to soak the soil, ocotillos leap into action, sprouting new leaves from base to tip, often within just a day or two. Have a look at one of those leaves if you can; delicate and thin, they are not the tough, waxy leaves of most desert plants. These leaves aren't here for a long time, just a good time. They work in overdrive, performing photosynthesis to store energy for the plant, but only as long as the water holds out. Once the soil is dry, the leaves fall off. It's a process that can happen multiple times every year. In some years, these drought-deciduous shrubs might grow and shed four or five sets of leaves.

All the energy that the plant races to produce goes into growing a beautiful spray of flame-red or orange flowers at the tips of those spiny stems that look like miniature torches. Though ocotillos can leaf out any time of the year, you're

Often mistaken for cacti, ocotillos are the US outliers in a small family of Seussian desert shrubs based almost entirely in Mexico.

most likely to see flowers in the spring and early summer. Flowers often remain on the plant after the leaves drop, giving you a decent chance of seeing flowers even after a dry spell. The lovely flowers produce nectar that is especially attractive to carpenter bees and hummingbirds—keep an eye out for both near blooming ocotillos.

One of the things that make BIBE so special is the wonderfully rich tracts of the Chihuahuan Desert that the park protects. Ocotillos are an intriguing and strangely beautiful part of that desert. Like many American families, they are cut off from their larger family by a human-drawn line, but are still an intrinsic and awesome part of life in BIBE and the desert Southwest.

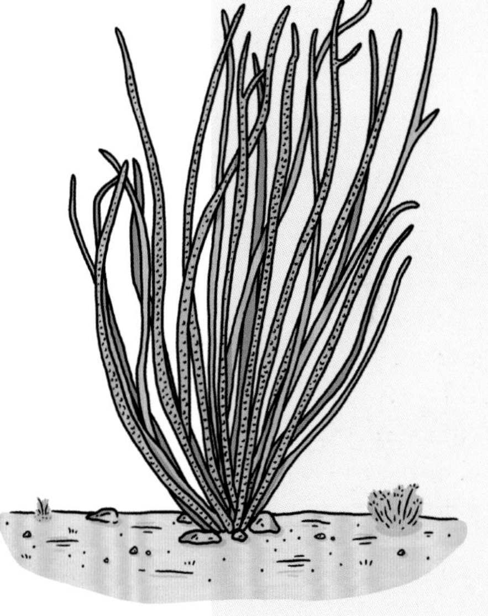

Ocotillo (*Fouquieria splendens*)

**Much of BIBE provides nearly unparalleled views of the Chihuahuan Desert and thus also of ocotillos. The plants are common and easy to identify.**

## DINOSAUR EXTINCTION: YOU'RE LOOKING AT IT

**SIXTY-FIVE MILLION YEARS** ago, an asteroid several miles across slammed into the sea near what is now the Yucatan Peninsula. Many scientists think that in the aftermath of the impact, hot debris engulfed the planet in flames, dust and dirt thrown into the atmosphere blocked the sun, and massive tsunamis churned across huge stretches of land. Some think this might not have been the only thing that wiped out the non-avian dinosaurs, but goodness, was it a blow! Certain places around the globe preserve a record of this catastrophe, including BIBE. Rocks from that time crop up in several places in the park, but a viewing tube at the Fossil Discovery Exhibit on Persimmon Gap Entrance Road (Highway 385) makes the boundary especially clear.

STATE
COLORADO

ANNUAL VISITATION
309,000

YEAR ESTABLISHED
1999

# black canyon of the gunnison

## sheer walls & shadows

| | |
|---|---|
| SUPERLATIVE | The Painted Wall is the tallest cliff in Colorado |
| CROWD-PLEASING HIKES | Cedar Point Nature Trail (easy); North Vista Trail to Exclamation Point (moderate); Oak Flat Loop Trail (strenuous) |
| NOTABLE ANIMALS | Mule deer (*Odocoileus hemionus*); yellow-bellied marmot (*Marmota flaviventris*); black-billed magpie (*Pica hudsonia*) |
| COMMON PLANTS | Utah juniper (*Juniperus osteosperma*; see page 134); needle-and-thread grass (*Hesperostipa comata*; see page 221); pinyon pine (*Pinus edulis*) |
| ICONIC EXPERIENCE | Craning your neck trying to find the river at the bottom of the canyon |
| IT'S WORTH NOTING | The Gunnison River drops 43 feet per mile through the canyon—that's more than five times as steep as the Colorado River's course through GRCA. |

**WHAT IS THE BLACK CANYON OF THE GUNNISON?** The Black Canyon is an unusually steep, narrow canyon along the (you guessed it) Gunnison River. It owes the "black" in its name to the dark metamorphic rocks of its walls and to being so narrow that some spots on the river below receive only minutes of sunlight each day. BLCA protects the deepest, most beguiling 14 miles of the chasm, less than 100 miles upstream from where the Gunnison's waters join the Colorado River and eventually flow through the Grand Canyon. Take a short hike along the rim, stare into the depths, and prepare to have your mind boggled by the immensity of the canyon and of time itself.

# GEOLOGY

## WORLD IN MOTION

The gneiss and schist of the canyon walls blend and change into each other because of small variations in the heat and pressure that altered the original rocks.

**WE EXIST ON A MOVING,** dynamic planet. The tectonic plates upon which our world is built are in constant motion. They move at about the same speed at which your fingernails grow, which seems tedious and plodding to us. But over unimaginably long stretches of time, this motion rewrites the globe's surface again and again. Scraps of ancient worlds survive, though. At BLCA, the canyon walls are made of ancient rock that is a powerful testament to the heat and restless motion of Earth.

The dark rock that makes up most of the canyon walls started out as seafloor sediments billions of years ago, at a time when life consisted of only single-celled algae and bacteria. These sediments—mud, sand, and volcanic ash—compacted into rocks without too much fuss. Had they stopped there, they might not look too different than many other rocks you find in this part of the country today. But instead, these rocks were buried deep underground and subjected to intense heat and pressure that cooked (metamorphosed) them into the hard, erosion-resistant schist and gneiss that makes up much of the canyon walls.

Cutting through the sheer rock faces are jets of light, pink-tinged rock. That, too, is ancient rock, though it is slightly younger than the darker rock. Sometime after the schist and gneiss had cooled and cemented, stress from the motion of plates fractured them. Like toothpaste being squeezed out

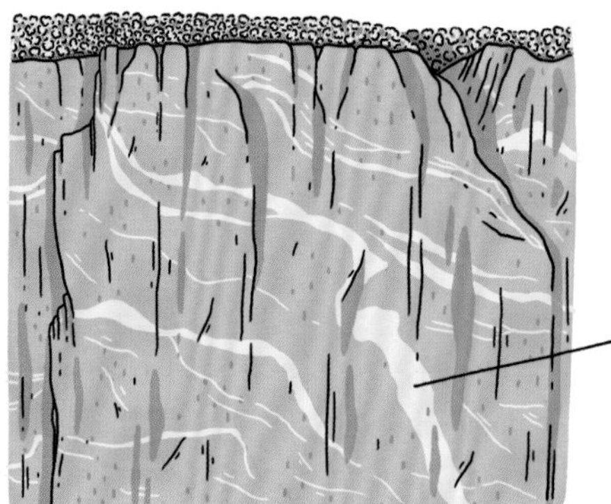

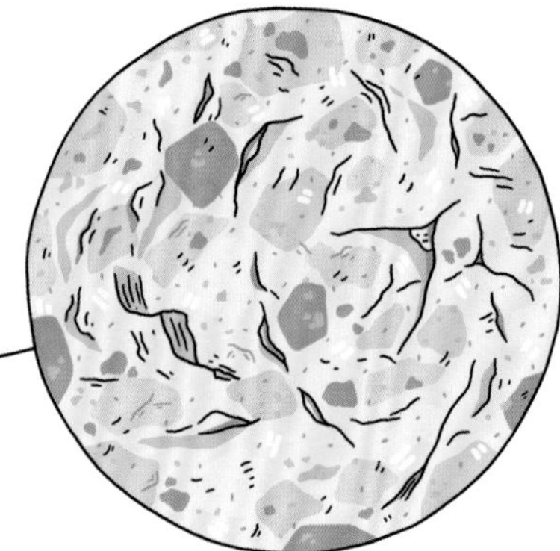

Pegmatite in the canyon walls cooled slowly, growing huge mineral crystals in the process.

of a tube, water-rich magma oozed through the fractures in the rock, leaving behind thick veins of pink pegmatite that crisscross the rock face. Look closely and you'll notice that some veins are neat and orderly and others look haphazard and chaotic. The varied patterns are a function of minute differences in the composition of the rock that the pegmatite intruded—change the rock and you change how it fractures and behaves under stress.

You can see pegmatite up close at a few overlooks that are perched on top of outcroppings. The blotches of color are mineral crystals—evidence that the magma inside the rock fractures cooled very slowly. The pink comes largely from a potassium mineral called feldspar, although sometimes it can look more yellow. Feldspar crystals easily reach 3 feet long in some places, and 6-foot-long crystals aren't unheard of. Look for smaller, but equally lovely quartz crystals; they tend to be clear or milky. With a little luck and some patience, you might even be able to find tiny garnets tucked in among the other minerals; they are usually more yellow than the red specimens you've likely encountered in jewelry.

You won't find many displays of ancient rock more stunning than the walls of the Black Canyon. The dark rocks tell stories about ancient seas from a time long before life as we know it. They also tell the story of how every rock on our planet is continually being formed, eroded, rearranged, and recombined. If the walls of the canyon look dynamic, it's because they *are*. You aren't imagining the movement they imply—you're reading it from across the eons.

**All of the canyon overlooks provide views of ancient schist and gneiss. Pink veins of pegmatite are also visible from just about any overlook, though they are more pronounced when viewed from the south rim. Several overlooks sit on top of pegmatite outcrops, including Gunnison Point.**

## and another thing

## PINYON PINES: ANCIENT TREES GROWING FROM ANCIENT ROCK

**THE GROVES OF** pinyon pines (*Pinus edulis*) adorning the canyon's rim are among the oldest yet found of their species. Tree-ring studies conducted in the 1940s and 1960s revealed that the pinyons growing near Warner Point were 750 to 850 years old at the time. Age doesn't always mean size, though—a short and stunted-looking pinyon might be centuries old. Looking near the base of the old-timers might reveal a glimpse of the next generation; young pinyons often grow in the shade and protection of older trees, biding their time until the older tree dies and they can take their rightful place in the sun.

STATE
TEXAS

ANNUAL VISITATION
172,000

YEAR ESTABLISHED
1972

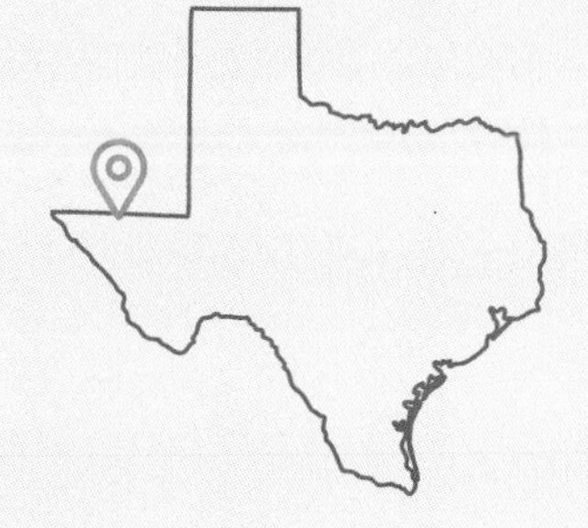

# guadalupe mountains

## an ancient reef flourishes in the desert

| | |
|---|---|
| SUPERLATIVE | GUMO contains the four highest peaks in Texas |
| CROWD-PLEASING HIKES | Manzanita Spring (easy); El Capitán Trail (moderate); Permian Reef Trail (strenuous) |
| NOTABLE ANIMALS | Mule deer (*Odocoileus hemionus*); Say's phoebe (*Sayornis saya*); rock squirrel (*Spermophilus variegatus*) |
| COMMON PLANTS | Honey mesquite (*Prosopis glandulosa*); prickly pear cactus (*Opuntia polyacantha*); ocotillo (*Fouquieria splendens*; see page 182) |
| ICONIC EXPERIENCE | Hiking or driving through McKittrick Canyon |
| IT'S WORTH NOTING | This is one park in which you will want to stay up late; GUMO has some of the darkest, starriest skies in the country. |

**WHAT ARE THE GUADALUPE MOUNTAINS?**
Hundreds of millions of years ago, massive colonies of algae and sponges built a limestone reef near the shore of an ancient sea. The Guadalupe Mountains are the eroded and uplifted remains of that reef—and the centerpiece of GUMO. Beautiful canyons cut into the sides of the mountains harbor wildlife and cooler temperatures that contrast sharply with the Chihuahuan Desert terrain that characterizes this area. Like all deserts, it is full of plants and animals carefully adapted to the rigors of living there. GUMO is a small park that packs incredible diversity into its landscape of mountains, canyons, ancient life, and desert specialists.

# GEOLOGY

## DEEP IN THE HEART OF THE PERMIAN

**THE WESTERN UNITED STATES IS** covered in all kinds of mountains: mountains made from plates crashing into each other (the Cascades), mountains made from bizarrely subducting plates (the Rockies), and mountains made from the stretching and thinning of continental crust (like, all of Nevada). But here in GUMO, the mountains are made from something else entirely. What look like chaotic textures and patterns are the fossils of untold numbers of sea creatures. These are not tectonic mountains; they are the remains of a massive reef. These rocks are a snapshot of time, frozen in place just before life on Earth changed forever in spectacular fashion.

During the Permian Period, 280 million to 250 million years ago, the continents were all smashed together in a landmass known as Pangaea. The land in GUMO was part of a broad basin just off the western shore. Over time, an enormous reef some 400 miles long grew out of the basin's floor. Just like modern reefs, this ancient structure was composed mostly of sea creatures and their calcium carbonate support structures, with large mats of algae holding it all together. Unlike modern reefs, the main players here weren't corals; they were sponges. Though you might think of sponges as squishy sacks of goo, some sponges grow hard internal structures that give them—and those around them—support.

The sponge *Amblysiphonella* is one of the most common fossils in GUMO. From the side, they resemble stacks of inner tubes.

All but one group of brachiopods were wiped out during the Permian Extinction. You can sometimes find large clusters of these two-shelled animals preserved in the rocks.

In the Permian, a diverse array of sponges formed the base of a reef every bit as dynamic as any in the modern world. Sharing the water was a cast of characters that included animals you might quickly recognize in a modern ocean, like sea urchins, horn corals, crinoids, and snail-like creatures. Those familiar sights were juxtaposed with altogether more fantastical beasts such as trilobites and scaphopods, a type of mollusk with a long, tapering, horn-shaped shell. All of their remains wound up jumbled and cemented together in the rocks that now make up the Guadalupe Mountains. As you hike, look for shell shapes, delicate lacy impressions, long tube forms, disembodied spikes, and even spots where it looks like individual grains of rice have been pressed into the surface of the rocks. These fossils, most the size of a quarter or even smaller, are a fantastic glimpse at a past world.

**Want to try your hand at identifying some of the fossils in these rocks? Copies of the aptly named Identification Guide to the Fossils of Guadalupe Mountains National Park can be downloaded from the park's website.**

Most of the animals immortalized in this reef have living relatives, but very few have survived unchanged through the vast stretch of time that separates then from now. Toward the end of the Permian, sea levels dropped, causing the water in the basin where this reef was growing to evaporate. Creatures that couldn't relocate died, but even those that could escape the dying reef could not escape the cataclysm that would shortly engulf the planet. The Permian mass extinction isn't as well known as the extinction that wiped out the non-avian dinosaurs at the end of the Cretaceous Period (or the mass extinction humans are currently causing), but by all accounts it should be. The Permian mass extinction, sometimes called the Great Dying, is the single largest mass extinction in the history of Earth. When all was said and done, roughly 96 percent of all species in the ocean had been wiped out. About 70 percent of all species on land vanished. What caused such an extreme event is a puzzle that is still being unraveled, but the world that it ended is preserved in the rocks of the Guadalupe Mountains.

Remember to never disturb any fossils you come across, and always include something familiar (like a lens cap, lip balm, or penny) for scale in your photos.

The reef was buried for eons until around 6 million years ago. Shifting tectonic plates once again uplifted and exposed it and its rich scenes of Permian sea life. You can see a long way from the top of the Guadalupe Mountains, but those paying attention on the way up can see not just far away, but far back in time as well.

**A variety of hikes in both the mountains and the canyons offer glimpses of this ancient reef. Published guides are available for some trails. Check in with a ranger for hikes that feature fossils and match your time and abilities.**

## and another thing

## FALL COLORS: EAT YOUR HEART OUT, NEW ENGLAND

**ENTERING ONE OF** GUMO's shady canyons is like walking into a different world than the desert that surrounds this place. Within the canyons, keep an eye out for deciduous trees such as bigtooth maples, gray oaks, and the Texas madrone, an evergreen with red bark that is unique to the region. Though they would be unable to survive in much of the park, these trees are perfectly content in the cool environs of GUMO's canyons. They even change color in the fall! Head out for an autumn canyon hike and you'll be treated to vibrant displays of reds, oranges, and yellows. Texas might not be known as a fall color destination, but anyone who has ever seen McKittrick Canyon decked out in its autumnal splendor knows better!

# great basin

**STATE**
NEVADA

**ANNUAL VISITATION**
153,000

**YEAR ESTABLISHED**
1986

## nevada's best-kept secret

**CROWD-PLEASING HIKES** Sky Islands Forest Trail (easy); Alpine Lakes Trail (moderate); Wheeler Peak Summit Trail (strenuous)

**NOTABLE ANIMALS** Red-winged blackbird (*Agelaius phoeniceus*); Great Basin rattlesnake (*Crotalus oreganus lutosus*); cave pseudoscorpion (*Microcreagris grandis*)

**COMMON PLANTS** Great Basin bristlecone pine (*Pinus longaeva*); fourwing saltbush (*Atriplex canescens*); big sagebrush (*Artemisia tridentata*; see page 123)

**ICONIC EXPERIENCE** Stargazing on a moonless night far from city lights

**IT'S WORTH NOTING** GRBA hosts an astronomy festival every autumn.

**WHAT IS GREAT BASIN?** The Great Basin is a region that stretches from Utah to California. It's actually a series of many basins separated by range after range of mountains. GRBA is arrayed around 13,063-foot Wheeler Peak, Nevada's second-highest peak. The park's main drag is Scenic Drive, a 12-mile road that ascends the side of Wheeler Peak, climbing 4,000 feet from desert to glaciers. There's animal life aplenty here; look in particular for changing bird populations as you gain elevation. With ancient trees, rock-covered glaciers, and a dazzling night sky, GRBA feels like a chunk of the Ice Age dropped into the middle of the desert.

## BOTANY

### PARTY LIKE IT'S 1999 (BCE)

**TREES CAN SEEM TIMELESS TO** us—we can spend our whole lives revisiting the same places, with old reliable trees standing tall as they always have. To us, a hundred years seems eternal. For trees, though, even several hundred years isn't all that extreme. Some trees can live more than a thousand years, and one species of tree outdoes them all with its longevity: the Great Basin bristlecone pine (*Pinus longaeva*).

Found at higher altitudes (most stands are at 9,000 to 11,500 feet) in the western United States, several Great Basin bristlecone pines have been found to be three thousand years old, with a select few coming in at *four thousand* years or older. The park and forest services keep locations of particularly ancient trees secret to protect them from vandals. All is not lost, though—in GRBA you can take a very doable 3-mile hike from Wheeler Peak Campground up to a grove of these ancient trees. There's even helpful signage once you get to the grove. Look for 1-inch needles growing densely from dangling branches. Because of their great age, they'll often have dead portions and heavily eroded bark as well.

So what's the deal with the long lives of these trees? It's really the intersection of some very special circumstances. First of all, they don't experience senescence, meaning they don't seem to experience adverse effects of aging like most other organisms do. Instead, they just keep living as they always have. One piece of evidence is that as the trees produce wood, their annual rings do not necessarily get narrower as they age and eventually

Great Basin bristlecone pine tree

become ancient, which is common among other trees. Age doesn't seem to slow these trees down as one would expect.

Ancient *Pinus longaeva* is also pretty good at selecting a place to call home; the harsher environment—with extreme winds, dry conditions, and frigid temperatures—decreases the likelihood of both being killed by fire (less kindlin') and being attacked by beetles, which aren't common at high altitudes. It's worth noting that the same trees living in more moderate climates simply grow taller, but don't live as long as their hardy, higher-altitude counterparts.

It's possible that offspring of long-living *Pinus longaeva* have "genetic memory" of past environs, meaning that they could have the potential to withstand climate change and ensure long-term survival.

All good things must come to an end, though, and since aging doesn't kill these trees, something else has to. Often, they die because the ground that has supported their roots for millennia erodes away. The tree falls, dies, and will eventually become soil to support future generations.

Great Basin bristlecone pine cone

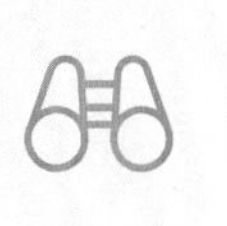

**Bristlecone pines can be found on Bristlecone Trail that begins at Wheeler Peak Campground. You might see the trees along other high-elevation trails in the park too.**

and another thing

## NIGHT SKIES: SEEING OUR COSMIC NEIGHBORHOOD

**GRBA'S NIGHT SKIES** are so dark that in addition to the Andromeda Galaxy (see page 138) you can often make out the fuzzy outline of the Triangulum Galaxy. Together, those two join our own Milky Way Galaxy as the three largest galaxies in the Local Group—a group of more than thirty of our closest galactic neighbors (mostly dwarf galaxies) all bound together by gravity. That means that on nights when the Andromeda and Triangulum Galaxies are visible along with the Milky Way (the only portion of our own galaxy that we can see), you are essentially standing on our galaxy's back porch looking out at our cosmic neighborhood.

STATE
**OHIO**

ANNUAL VISITATION
**2.1 MILLION**

YEAR ESTABLISHED
**2000**

## an urban wilderness in recovery

| | |
|---|---|
| CROWD-PLEASING HIKES | Blue Hen Falls Trail (easy); Boston Run Trail (moderate); Towpath Trail (strenuous) |
| NOTABLE ANIMALS | Raccoon (*Procyon lotor*); bullfrog (*Rana catesbeiana*); great blue heron (*Ardea herodias*) |
| COMMON PLANTS | Red oak (*Quercus rubra*); New England aster (*Symphyotrichum novae-angliae*); Canada goldenrod (*Solidago canadensis*) |
| ICONIC EXPERIENCE | Visiting Brandywine Falls |
| IT'S WORTH NOTING | You can visit the Countryside Farmers' Market at Howe Meadow from April to October—some of the items sold there are grown within the park! |

**WHAT IS CUYAHOGA VALLEY?** Located near several urban areas in Ohio, the central feature of the Cuyahoga Valley is the Cuyahoga River—a waterway best known for having been on fire a startling *thirteen* times since the mid-1800s. Its 1969 fire sparked parts of the environmental movement and spurred establishment of the Clean Water Act (1972). CUVA was designated a National Recreation Area during the cleanup efforts of the 1970s and then made a national park in 2000. It doesn't represent wilderness in the way that many other national parks do—after all, this is the only wilderness park designated on the site of an environmental disaster—but the park is an ongoing example of a space regenerating from human disturbance.

## WILDLIFE

### BEAVERS RECLAIM THE SALVAGE YARD

**HUMANS HAVE LONG REFERRED TO** beavers (*Castor canadensis*) as "nature's engineers" because they are so skilled at building their elaborate underwater dams and lodges. In CUVA, they're celebrated as the most determined rescue-engineers of the park. CUVA's environmental crises weren't limited to dirty river water. After the park's establishment as a National Recreation Area in 1974, the NPS started purchasing private land in the Cuyahoga Valley to ensure that the region would be protected for future generations to enjoy. Often, those properties were home to businesses that negatively affected the land. One of those properties was an auto repair shop and salvage yard. The buildings were demolished in the 1980s, but debris and pollutants continued to linger in the area. The NPS considered turning the new land into a parking lot, but some beavers moved in before that could happen. They then dammed surrounding waterways, flooding the area. Beavers had previously been extirpated from Ohio in the 1800s by fur trappers, so their return was a significant ecological win for the area. Volunteers used canoes to clean up debris in the newly formed wetland, and the beavers made it clear that they'd keep flooding the land if humans kept trying to build on it.

That former salvage yard is now known as Beaver Marsh. It's overflowing with plants and wildlife that can be viewed year-round. The National Audubon Society has designated CUVA as an Important Bird Area, worthy of protection due to its avian populations. With more than two hundred species of birds known to frequent the park, there are excellent opportunities for birdwatching. Beaver Marsh is especially rich during the annual spring and fall migrations. Beginning in March, look for great blue herons (*Ardea herodias*), red-winged blackbirds (*Agelaius phoeniceus*), and tree swallows (*Tachycineta bicolor*). Early spring brings birds returning from their wintering grounds down south, while year-round residents, like bald eagles (*Haliaeetus leucocephalus*), might be out and about in the warming weather. In November, you might see wood ducks (*Aix sponsa*) and white-throated

## CREATURES OF BEAVER MARSH

Keep your eyes peeled during your visit—you can see wildlife everywhere, including at your feet! Here are just a few examples of the creatures you might come across.

**Red-winged blackbird (*Agelaius phoeniceus*)** A small, stocky bird. Males are black with a red shoulder patch (hence the name); females are a streaky brown. They often appear hump-backed when perched in a tree.

**Great blue heron (*Ardea herodias*)** A large bird (up to 4 feet tall) with blue-gray coloration. They have a dagger-like beak and long legs, and you're likely to see them wading in the marsh.

**Muskrat (*Ondatra zibethicus*)** A dark-brown to black mammal, 1½ to 2 feet long. They look similar to beavers, but are smaller and swim with their entire bodies visible. (Typically only beavers' heads are visible when they swim.)

**American bullfrog (*Rana catesbeiana*)** A large (3½- to 6-inch-long) green or gray-brown frog. Listen for the males' bellow, which sounds like a cow's moo.

sparrows (*Bombycilla cedrorum*) en masse as they're making their way south. As is common with wildlife viewing, the best time to see most bird species is early morning and evening.

Birds are only part of the wildlife story of Beaver Marsh, though. It's possible that you'll see beavers swimming by with branches of willow or aspen, setting off to work on their lodges. You may also see snapping turtles (*Chelydra serpentina*; see page 276) swimming among the abundant lily pads. One variety of this aquatic plant is spatterdock (*Nuphar lutea*)—look for yellow, half-opened flowers among its massive leaves. In the last decade, river otters (*Lontra canadensis*) have set up camp in the marsh—a significant win for the area, since otters are water-quality snobs and won't tolerate pollution. Other parts of the marsh are still regenerating though—years of environmental damage aren't overcome quickly.

To see a wetland that is still in the very early phases of regeneration, pay a visit to the Krejci Dump Site inside the park. The NPS took control of the former dump in 1985 and initially assumed the cleanup wouldn't be too intensive. It turned out that the dump's waste was toxic and entirely too dangerous for a simple cleanup—substances such as arsenic, herbicides, paint, and any number of heavy metals and other chemicals had to be removed. The Environmental Protection Agency (EPA) initiated emergency cleanup efforts, but because of the scale and cost of the project, legal action was taken against parties responsible for the dumping. Even though the dumping was legal at the time, there are laws holding parties financially responsible for environmental damage caused by their dumping. Eventually, Ford and GM settled, along with other companies, and Ford spearheaded the cleanup with financial help from GM. Intense cleanup efforts concluded in 2012, followed by two years of revegetation of the entire site. What was once a disaster zone is now a place of regeneration and new life. The site certainly isn't as lush as Beaver Marsh, but it's hoped that as plant and animal life continue to rebound, this wetland will become another excellent example of nature's ability to heal itself.

**Accessed by Towpath Trail, a short paved trail and boardwalk lead to Beaver Marsh. The Krejci Dump Site is located near the Boston Store Visitor Center along Hines Hill Road, but there's no formal parking lot, so you'll need to park on the roadside or hike in.**

# indiana dunes

STATE
**INDIANA**

ANNUAL VISITATION
**1.8 MILLION**

YEAR ESTABLISHED
**2019**

## glacial sand dunes amid steel mills

| | |
|---|---|
| SUPERLATIVE | INDU is the only park designated in 2019 |
| CROWD-PLEASING HIKES | Portage Lakefront and Riverwalk (easy); Tolleston Dunes Trails (moderate); West Beach Trail (strenuous) |
| NOTABLE ANIMALS | White-tailed deer (*Odocoileus virginianus*); sandhill crane (*Antigone canadensis*); beaver (*Castor canadensis*) |
| COMMON PLANTS | Sphagnum moss (*Sphagnum* spp.); black oak tree (*Quercus velutina*); beach grass (*Ammophila breviligulata*) |
| ICONIC EXPERIENCE | Sunbathing on the shores of Lake Michigan |
| IT'S WORTH NOTING | INDU is internationally recognized for its year-round birding, but especially notable are periods of migration when it's possible to see a hundred species of birds in a single day! |

**WHAT ARE INDIANA DUNES?** Rising almost 200 feet above the electric-blue waters of Lake Michigan, the dunes are massive piles of sand created by the same sheets of ice that carved and filled the Great Lakes. INDU protects a 15-mile stretch of wild glacial lakefront along the southeastern shore. Marching backward from the water are a succession of dunes, the youngest of these sprouting grasses, the oldest now crowned with prairies and forests. Steel mills and the Chicago skyline are the surprising backdrop for the rich diversity of bird life found here—between permanent residents and migratory species, it's a year-round party. Don't forget to pack your binoculars and your bathing suit!

## GEOLOGY

### FEELING THE PULL OF THE SAND

The abundance of quartz grains is what makes the sand "sing" when you walk on the dunes—it's the sound of glassy quartz grains rubbing against one another.

**STROLLING THE SANDS AT INDU,** you're likely to notice dark streaks among the lighter grains. Some people assume that these streaks are made of dirt or maybe pollution of some kind—there's an awful lot of industry on Lake Michigan, after all. But it's neither of those—in fact, the black grains are actually a kind of magnetic sand!

Made of magnetite and ilmenite, these black sand grains are tiny pieces of iron. They are small fragments of iron-rich rocks from the Upper Peninsula of Michigan—a huge reason for all that industry on the Great Lakes in general. The iron up there is more than a billion years old and contains some of the first evidence of oxygen in early Earth's atmosphere. Much later, beginning around 2 million years ago, massive glaciers from the north pulverized parts of those rocks and carried them south to the present-day shoreline of Lake Michigan. As wind and water continuously sort sand grains on the lakeshore, the dark grains, which are heavier than the quartz that makes up most of INDU's sand, end up collected together in streaks and patches. Near the waterline it can look like an oily black smudge; in drier locations, the grains can look like contour lines drawn on light sand.

You don't have to wait for the elements to do their work. Go ahead and scoop up a handful of dry sand—tucked in among the tan grains of quartz are specks of magnetite. You can see them without a loupe or magnifying glass, but grab a magnifier anyway because looking at magnified sand is always interesting. The real show here requires one more piece of super-simple equipment: a magnet. One from your fridge or the gift shop will do just fine. Find a nice streak of black and watch the grains leap up to your magnet!

Magnetite is one of a very small number of naturally magnetic minerals and the only one that is fairly common. Scattered through sand dunes, magnetite is a cool distraction. But inside its source rocks, it takes on an important role as storyteller about Earth's past. When rocks form that contain magnetite, the crystals orient themselves along the planet's magnetic field. Because this field reverses itself at irregular intervals, tracking the orientation of magnetite grains in layers of rocks sketches out a map of not only Earth's past magnetism, but also the continuous movement of tectonic plates. It's a pretty cool story!

Watch the dark grains leap toward the magnet in your hand. Those are pieces of our past planet—fragments of a much, much longer story. If any of your gear, like a water bottle clasp or compass, is magnetic, you'll quickly note that these scraps of the past are also very difficult to keep off the magnet. Enjoy these ancient grains and their stories while you explore the dunes!

**Always stay on designated trails when exploring the dunes to avoid damaging the dunes or their specialized plant life.**

**Magnetite can be found in most of INDU's sand, both on the beaches and in the dunes.**

## BOTANY

### A STABILIZING INFLUENCE

**AS ANYONE WHO HAS EVER** tried to clean sand out of shoe treads or gotten stung by its grains on a windy day can attest, sand can be a rather insidious and pernicious substance. Blowing around in the wind is one of sand's favorite activities. Retaining nutrients or water is much lower on the list of things sand does willingly. A constantly shifting, nutrient-poor substrate is far from what most plants need to survive, but up from this sand rises a champion: beach grass.

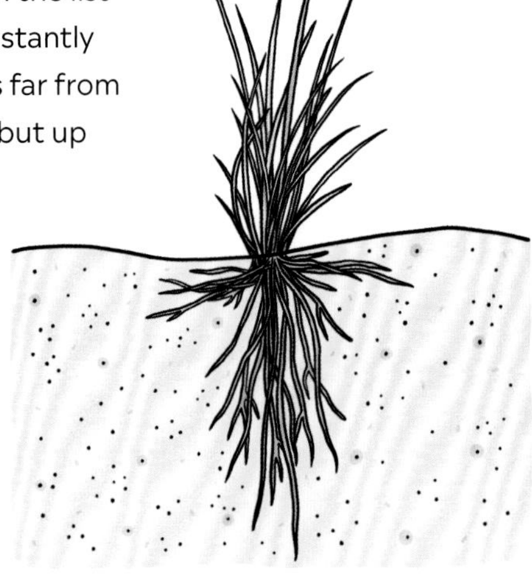

Beach grass (*Ammophila breviligulata*)

Also called marram grass, beach grass (*Ammophila breviligulata*) is a hugely successful colonizer of sand dunes. Just about the only plant on young dunes, it is found all over the Great Lakes as well as up and down the Atlantic Coast, from Newfoundland to North Carolina. Get up close and have a look at some. Narrow green leaves grow directly out of the sand and can be a few feet tall. Run your finger gently across the leaves and notice the rough top side and smoother underside. In summer, look for the plant's seed head: a tall, subtly pyramid-shaped spike of densely clustered beige flowers.

The seed head and leaves are just the tip of this vegetative iceberg. Below the sand, each bunch of beach grass is a sprawling, multilayered network of rhizomes. These stemlike roots extend deeply downward and outward in all directions from the plant. Downward rhizomes provide stability and access to water; horizontal rhizomes send up new shoots that become new clumps of beach grass. These rhizomes can reach more than 20 feet in length, an extensive network that provides beach grass with the stability it needs to thrive in this shifting world of sand.

For many plants, a mantle of sand would spell the end. For beach grass, though, burial by sand is just another day and another opportunity. Rather than shrinking from it, beach grass embraces burial. As sand builds, the stems simply grow taller and the rhizomes get deeper.

Look up at the towering dunes. Clumps of beach-grass leaves act as a windbreak for sand particles, while below ground, huge rhizome networks stabilize the dune from within and help prevent erosion. These stacks of sand are being held together by beach grass!

As the grass grows and the dunes stabilize, the leaves and seed heads trap not just sand but also organic debris blowing in the wind. This debris, and the plant's own fallen leaves and decomposing roots, greatly enrich the sand with otherwise absent nutrients. By fostering more stable sand and amplifying available nutrients, beach grass readies the dunes for the advance of other plants. Other grasses arrive first, followed by wildflowers and shrubs. Eventually, after 2,000 years or so, the dunes are stable and rich enough to support entire forests of black oak, hickory, and sassafras trees.

At INDU it's possible to hike through dunes in all phases of plant colonization. Watch as waving grasses besieged by sand are replaced by blankets of green and then forests of trees. It's like watching a fascinating time lapse of thousands of years of storms, sand, and survival—all of which is deeply rooted (pun intended) in the tenacity and hardiness of humble beach grass.

Cottonwood trees (*Populus deltoides*) are another early dune colonizer. They grow in more sheltered spots, but can also survive sand burial. When buried, the branches begin to act like roots. A cottonwood that appears short may be the tip of a tree more than 50 feet tall.

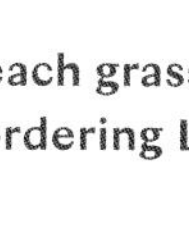

**Beach grass grows along the dunes bordering Lake Michigan.**

## MIGRATING BIRDS' INTERNAL COMPASS

**A HUGE NUMBER** of bird species, such as sandhill cranes and Kirtland's warblers, stop along Lake Michigan's shores and the surrounding wetlands to rest and refuel during their annual migrations. How exactly birds navigate while traveling thousands of miles is still not fully understood. They likely use a variety of techniques, including visual landmarks and the orientation of the sun and stars. The other key component is likely the avian ability to sense Earth's magnetic field. Particles of magnetite in bird beaks and specialized magnetically sensitive eye proteins seem to play a role, but so far no one has figured out exactly how information from these internal magnetic sensors translates to actual navigation.

STATE
ARKANSAS

ANNUAL VISITATION
1.5 MILLION

YEAR ESTABLISHED
1921

# hot springs

## hot water & history

| | |
|---|---|
| SUPERLATIVE | HOSP was the first piece of land set aside for federal protection |
| CROWD-PLEASING HIKES | Grand Promenade (easy); West Mountain Trail (moderate); Sunset Trail (strenuous) |
| NOTABLE ANIMALS | Eastern chipmunk (*Tamias striatus*); pileated woodpecker (*Dryocopus pileatus*); five species of ticks |
| COMMON PLANTS | Oak trees (*Quercus* spp.); wild bergamot (*Monarda fistulosa*); cyanobacteria (*Phormidium treleasei*) |
| ICONIC EXPERIENCE | Soaking in the mineral waters |
| IT'S WORTH NOTING | You can get a bird's-eye view of the rolling mountains around the park from the top of Hot Springs Mountain Tower. |

**WHAT ARE HOT SPRINGS?** Hot springs are springs of naturally warm water that humans have been soaking in for millennia. Small, urban HOSP is a different kind of national park. Its dozens of naturally occurring thermal springs were given federal protection in 1832, predating the establishment of the National Park Service by more than eighty years. (HOSP was not designated a national park until later, meaning that YELL is indeed the world's first national park.) The main drag of HOSP is a row of historic bathhouses offering traditional baths, museum exhibits, and shops. Hiking trails wind through the low mountains surrounding Bathhouse Row, offering a quick escape from the crowds. Spring water is freely available throughout the park. So grab a cup, slip into some flip-flops, and get ready to take in a park like no other.

# GEOLOGY

## HOT WATER, HOLD THE VOLCANO

**SPEND A MINUTE STUDYING THE** map of where thermal springs pop up in the continental United States on the facing page and you stand a good chance of noticing something odd about the location of Hot Springs, Arkansas.

On the map, a line of thermal springs runs down the spine of the Appalachians, and a whole mess of them crop up in the West. Those are all places that are volcanically active now or were in the past, so those springs make sense. But what about the ones in Arkansas? The most recent igneous rocks in this state are 90 million to 100 million years old. They haven't been giving off heat for a while.

When you take a soak in a HOSP bathhouse or have a refreshing sip of that warm water, you are experiencing 4,400-year-old rainwater that has been slowly heated deep underground. Here comes the rain again!

Thousands of years ago, rainwater fell on mountains northeast of the modern park border. The mountains in this region are made of brittle rock that was roughed up during mountain building. You can see it in outcrops around the visitor center and along trails—the rocks look every bit as cracked and fractured as they are. All those cracks let water percolate down through the rocky crust of our planet.

Coal miners and geologists know that temperatures rise as you go deeper underground. The heat is actually left over from the formation of this planet more than 4 billion years ago. Particles, moons, asteroids, and other rocky bodies slammed and merged into the earth, building an ever-larger rocky planet, and also generating immense amounts of heat. That heat spreads outward in convection currents from our planet's molten core into the cooler parts of the mantle and the crust. Adding to Earth's interior sauna are slowly decaying radioactive elements, like uranium, thorium, and potassium. Their decay releases energy in the form of heat. So even after a few billion years, our planet is still cooking from the inside out.

As rain fell on the nearby mountains thousands of years ago, it percolated through the rocks and came into contact with our planet's inner heat. The rainwater traveled happily along underground for thousands of years, being slowly heated up before being rapidly forced to the surface through a series of faults just to the west of Bathhouse Row—arriving so quickly, in probably less than a year, that it's still hot enough to ease aching bones and muscles. Think about that while you soak in a bathhouse. When that water fell as rain, Indigenous

## UNITED STATES OF THERMAL WATER

The continental United States boasts many thermal springs, most of which are volcanically heated. Tiny HOSP is the exception; its waters are heated by the lingering heat of our planet's formation and the slow decay of radioactive elements underground.

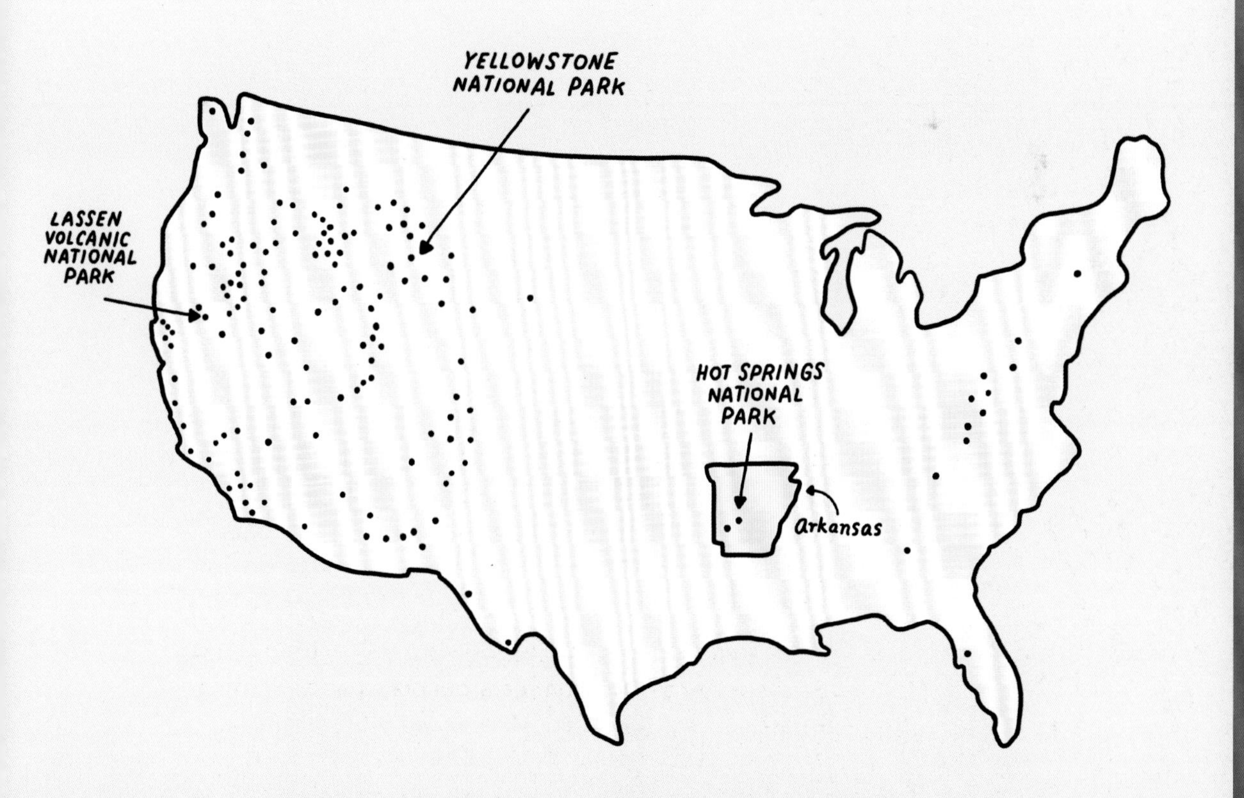

people in the Americas were busy domesticating corn, the Sumerians were writing things down for the first time, and northern Europeans were making things out of bronze with which to hit each other.

Today, you can soak in prehistoric water that's been pumped into a bathhouse pool or even see it burbling directly out of the ground along the Promenade. However you decide to explore the park, know that this is one kind of hot water that it's good to get yourself into!

**The Buckstaff Bathhouse and Quapaw Baths and Spa both offer visitors the chance to soak in thermal waters. Check websites for pricing and services. Most of the natural springs in HOSP are capped for safety reasons, but you can see them along the Promenade. To see a free-flowing spring, check out the "Hot Water Cascade" at the far north end of the Promenade.**

STATE
SOUTH DAKOTA

ANNUAL VISITATION
1 MILLION

YEAR ESTABLISHED
1978

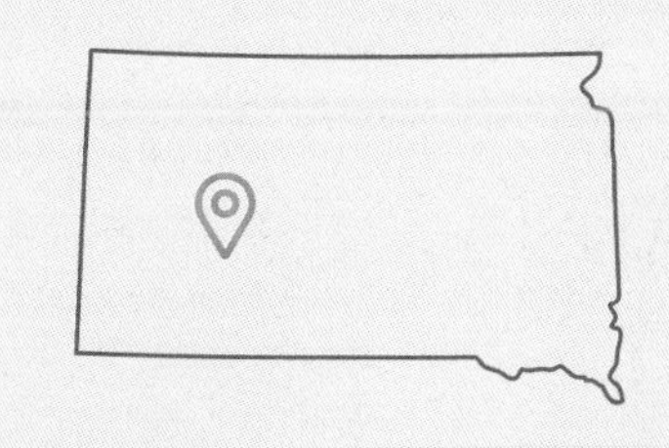

# badlands

## grasslands give way to colorful buttes

| | |
|---|---|
| SUPERLATIVE | BADL contains one of the richest assemblies of fossil mammals in the world |
| CROWD-PLEASING HIKES | Fossil Exhibit Trail (easy); Medicine Root Trail (moderate); Notch Trail (strenuous) |
| NOTABLE ANIMALS | Bison (*Bison bison*); black-tailed prairie dog (*Cynomys ludovicianus*; see page 227); prairie rattlesnake (*Crotalus viridis*) |
| COMMON PLANTS | Western wheatgrass (*Pascopyrum smithii*); upright prairie coneflower (*Ratibida columnifera*); prickly pear cactus (*Opuntia polyacantha*) |
| ICONIC EXPERIENCE | Watching the buttes at sunset |
| IT'S WORTH NOTING | The park is divided into two distinct units. The North Unit contains many visitor services. The Stronghold (South) Unit is largely undeveloped and has few paved roads—check in with a ranger before making plans to visit that portion of the park. |

**WHAT ARE BADLANDS?** Badlands are arid regions of land where soft rock is eroded into strange shapes by wind and water. The rugged terrain—prone to crumbling and collapsing when dry, and softening into an impassable clay when wet—and lack of drinkable water make them "bad" to travel across. Badlands occur all over the world, but they are very much associated with the Great Plains of the United States. BADL protects one such stretch of wild and twisted terrain. As the soft, colorfully banded formations of BADL erode, they reveal one of the richest assemblages of fossils in world—affording those keen enough to look a front-row seat for the rise of the mammals.

## GEOLOGY

### MAMMALIAN YEARBOOK

**OF ALL THE BILLIONS OF** animal species that have ever lived on this planet, only an infinitesimal fraction have been preserved in the rocks as fossils. But every so often there's a time and a place where all the conditions are right to give us a startlingly robust picture of an ancient landscape. BADL contains one such magical slice of time. Millions of years ago, conditions here were just right to preserve creatures in the fossil record, and today conditions are just right to expose those fossils. And the tales those fossils tell is one of a world increasingly dominated by creatures like us: mammals. The rocks here are like a yearbook of the mammalian class between 37 million and 30 million years ago—portions of the Eocene and Oligocene epochs. And like any good yearbook, these rocks capture not just the cast of characters but the events in the world around them as well.

The Eocene and Oligocene epochs were an important time in the history of our planet, the first moment that Earth as we know it was taking shape. The continents were heading toward their current locales. Their movement shifted ocean currents and cooled the planet's climate way down. This era was when mammalian life began to flourish as never before.

Mammals have been around for at least 200 million years, but the rocks exposed in the park represent a particularly rich time for us. When the non-avian dinosaurs went extinct around 65 million years ago, they left many niches for life to fill. By the Eocene, 37 million years ago, mammals had shouted "Don't mind if we do!" and were in the midst of an evolutionary party as if their parents were away on vacation.

The fossilized evidence of that party can be found all over BADL in gray, tan, and red rock layers. Gray layers largely represent the Eocene-aged Chadron Formation. Between 37 million and 34 million years ago, those gray rocks were the soil of a subtropical forest with a climate similar to that of present-day South Carolina. Running through that forest were adorable dog-size horses, one of the few native North

## FOSSIL SPOTTING 101

Learning to spot fossils takes practice, but it also requires a serious commitment to take only photos. Moving or removing fossils robs them of their original context; it is rude and illegal. You are free to wander off-trail in BADL, but take care not to trample plants or disturb fossils.

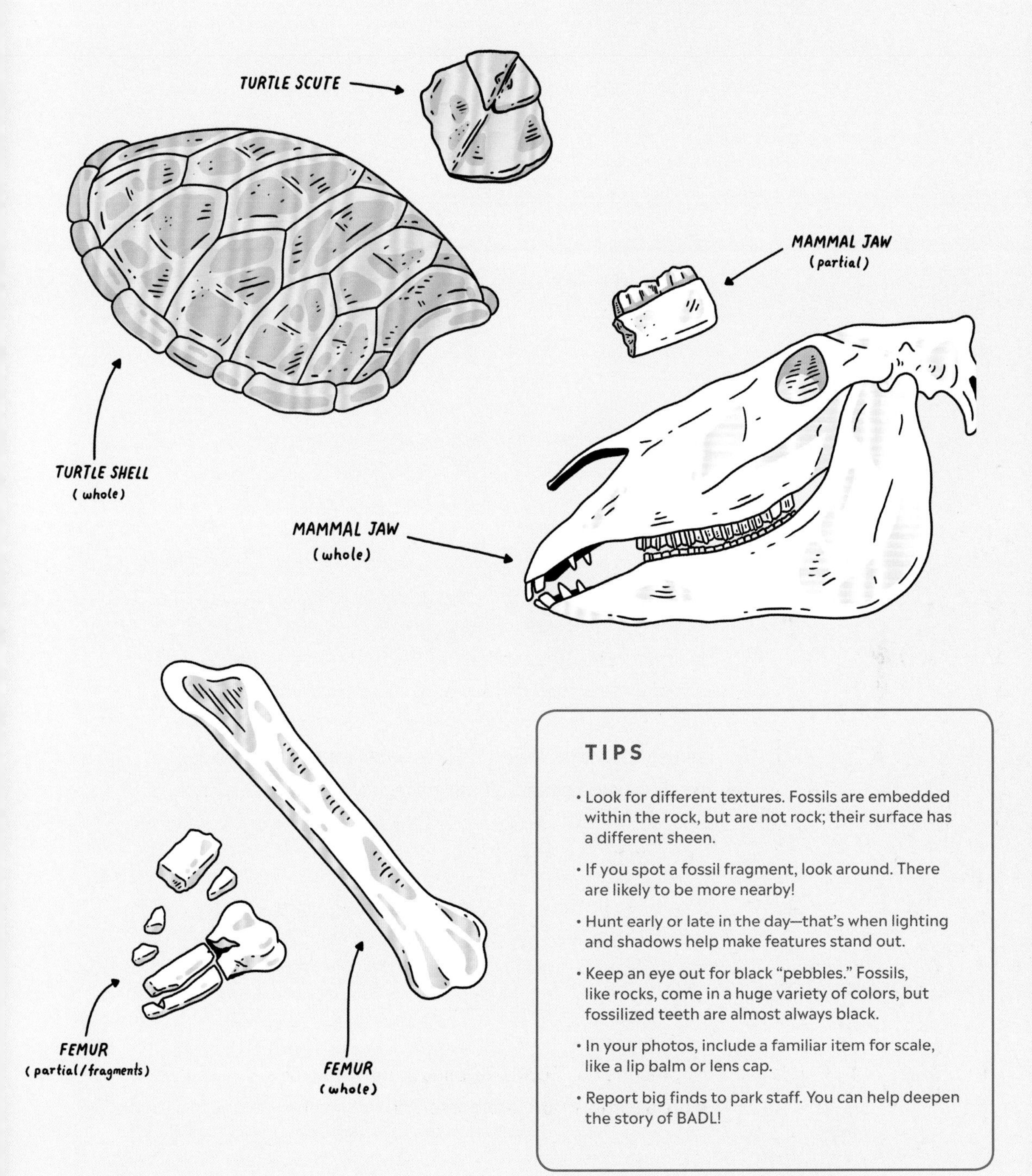


### TIPS

- Look for different textures. Fossils are embedded within the rock, but are not rock; their surface has a different sheen.
- If you spot a fossil fragment, look around. There are likely to be more nearby!
- Hunt early or late in the day—that's when lighting and shadows help make features stand out.
- Keep an eye out for black "pebbles." Fossils, like rocks, come in a huge variety of colors, but fossilized teeth are almost always black.
- In your photos, include a familiar item for scale, like a lip balm or lens cap.
- Report big finds to park staff. You can help deepen the story of BADL!

American animals. (Many of the species we think of as our own actually evolved elsewhere and migrated to our humble continent.) Their teeth are among the most common fossils found in BADL. Some of the most remarkable fossils from this layer come from titanotheres, which looked like overgrown rhinos with forked horns and were the largest beasts on the North American landscape.

The majority of Chadron Formation fossils are found in the more remote southern portion of the park. Visiting there requires permits and a sturdy vehicle. See a ranger for details.

About 34 million years ago, the climate in this area became increasingly drier and cooler, and the forest gave way to an open savannah crisscrossed by rivers and streams. You can see remnants of this time in the tannish brown-and-red-banded rocks of the Brule Formation. Inhabiting the braided river channels were many kinds of turtles. Keep an eye out for their scutes—individual shell plates—which are plentiful as fossils. More rough and tumble are the fossils of *Archaeotherium*—extinct pigs who, judging from the healed injuries on many of their fossils, got into a lot of fights with each other.

The sheer abundance of fossils from this slice of time allows us to reconstruct entire food webs and landscapes, rather than isolated creatures. We don't have 7 million years to watch modern climate change unfurl and alter landscapes and animals, but we can read that story in the rocks of BADL. As you walk around, remember that it's possible to put together all those little mammal bits you're seeing into one very big storyline.

**There are some especially accessible exposures of the Brule Formation in the Cedar Pass area. The Ben Reifel Visitor Center houses great exhibits about these fossils and offers the chance to peek into a working fossil-prep lab. Exploring outcroppings of the Chadron Formation in the Stronghold (South) Unit requires permission and preparation.**

## BOTANY

### IT'S A JUNGLE OUT THERE

**THE LANDSCAPE IN BADL LOOKS** otherworldly. Harsh climate, unstable land, and a limited amount of rainfall mean there aren't many trees here; open prairie dominates. This current prairie is just another in a long series of ecosystems that this land has hosted over millions of years. Lush forests once grew here, and the evidence is found throughout BADL in the saturated tones of yellow, orange, and red on rounded mounds amid the more sedately colored pinnacles and spires. Those bright colors are the fossilized soils, or paleosols, of a long-ago rich, forested land.

Pinnacles and spires

Look first at the yellow and orange layers. They date from the early Eocene epoch, about the time that the Rocky Mountains were rumbling upward and shortly after the non-avian dinosaurs made their exit. At that time, a once-vast inland sea was draining out of this area, leaving behind a thick, nutrient-rich muck. Brackish pools and streams formed in the hilly landscape; and where the muck was exposed to the elements, it weathered into a fertile soil.

Mounds

Trees—some probably quite large—grew up from the newly weathered soil. These layers of ancient soils record their growth. When we think of fossils, most of us think of those that preserve an actual body or plant part—scientists call those *body fossils*. But there are other fossilized records of life, such as insect burrows and the root cavities of plants. Those are the fossils that this jungle left behind. They are difficult to spot—some of them are no more than slightly dark staining around a former root—but they speak volumes about a landscape. After all, the root profile of a thick forest looks very different from the much shallower root systems of a savannah. For those of us less trained in paleontology, the bright colors will do just fine as our jungle ID.

The red and purple layers above the yellow and orange layers are slightly younger and represent a time when brackish pools were replaced by a network of freshwater streams and rivers moving across a plain. In this case, the color itself is a clue that points to changing conditions. Red paleosols are indicative of well-drained and aerated conditions that allow lots of oxygen into the soil, where it reacts (rusts) with iron minerals.

Notice that the brightly colored paleosol layers erode into gently rounded "haystacks," very much in contrast to the pinnacles and buttes of the park's younger layers. These soils were not always these colors; there were no jungles springing out of yellow soil. The bright colors developed after the soils were buried and turned to rock, but they preserve important information about the deep history of this place. They are a rainbow time machine transporting us to those ancient forests.

**Exposures of these paleosols crop up along several spots on the main Loop Road, including Yellow Mounds and Sage Creek Overlook. They are not difficult to spot. For a look at nearly the whole sequence of Badlands rocks from oldest to youngest, check out Dillion Pass.**

STATE
NORTH DAKOTA

ANNUAL VISITATION
749,000

YEAR ESTABLISHED
1978

# theodore roosevelt

## rugged badlands & rolling prairies

| | |
|---|---|
| SUPERLATIVE | THRO is the only national park named for a person |
| CROWD-PLEASING HIKES | Coal Seam Trail (easy); Sperati Point (moderate); Caprock Coulee (strenuous) |
| NOTABLE ANIMALS | Bison (*Bison bison*); pronghorn (*Antilocapra americana*; see page 121); feral horse (*Equus caballus*) |
| COMMON PLANTS | Rocky Mountain juniper (*Juniperus scopulorum*; see page 134); sagebrush (*Artemisia* spp.; see page 123); prairie rose (*Rosa arkansana*) |
| ICONIC EXPERIENCE | Watching wildlife on the grasslands |
| IT'S WORTH NOTING | Though the management of the herd has not been without controversy, THRO is one of the few places on the plains where you can easily see free-roaming horses. The herd is small enough that an ID guide to the individual horses is published every year. |

**WHAT IS THEODORE ROOSEVELT?** Theodore Roosevelt the human (1858–1919) was the twenty-sixth president of the United States. Theodore Roosevelt the national park is a prairie grassland dappled with colorfully striped badlands and home to iconic animals such as bison and feral horses. The park is actually parceled into three distinct units separated by many miles and a time zone. The North (central time) and South (mountain time) Units feature scenic drives, interpretive trails, hikes, and ranger programs. The much smaller Elkhorn Ranch Unit preserves the site of Roosevelt's former cabin, though the building itself has been moved to the South Unit. Seemingly infinite prairie vistas coupled with a good amount of solitude await explorers here.

STAMP
PASSPORT
HERE

# GEOLOGY

## VOLCANIC BLUES

The term *badlands* refers to more than just the national park of that name. It is also any terrain made of quickly eroding rocks and soil, with little vegetation, and virtually no drinkable water.

**ASK ANY NUMBER OF PEOPLE** where they might go to see volcanoes in the United States and "the Great Plains" is probably not an answer you'll hear very often. But sure enough, there is evidence of massive volcanic eruptions running all through THRO in plain sight.

Between 70 million and 40 million years ago, while the Atlantic Ocean was opening along the East Coast, one of Earth's massive tectonic plates was sliding under the North American plate on the West Coast. As it did, huge peaks rose far inland from the collision zone: the Rocky Mountains. As they rose, magma from beneath the plate boiled up through the crumpled rock and erupted. Widespread volcanism poured out huge amounts of lava, accompanied by great quantities of ash.

Wind and water spread the ash out across the northern Great Plains, with quite a bit of it coming to rest on what is now THRO. At the time, this land hosted winding rivers, forested floodplains, and vast swamps. Imagine the Everglades in Florida but with 100-foot-tall trees with trunks 12 feet around. Fossilized leaf imprints and pollen from the area show evidence of cypress and sequoia trees, tree ferns, and many other plants that vanished from this area long ago.

As the ash settled in the water-logged landscape, it slowly weathered into a soft clay called *bentonite*. This ash-turned-clay makes up striking blue bands that can be traced for miles within the striped rocks of the park's badlands.

When dry, the blue clay cracks into a characteristic texture called *popcorn*. When wet, it unleashes its wild side. Its grains are capable of absorbing enormous amounts of water and swelling to many times their original size. The muck that results is infamously sticky and slippery at the same time. The locals call it "gumbo clay." Human feet, horse hooves, wooden wheels, and modern tires are no match for bentonitic mud.

When wet and on a slope, bentonite will flow downhill. You can see it in a few different ways. In places where bentonite layers are near the top of a rock formation, the clay spills downward, making the rest of the formation look like someone has spilled a bucket of chalky blue-gray paint down its front. When bentonite layers in the middle of a formation give way, entire blocks of earth slip downward like the layers of a cake slipping off one another. Geologists call this *slumping*. Look for places where the normally horizontal bands of rock are tilted at an angle.

In addition to blue-striped badlands, the ash left one more gift: petrified wood. Silica, a natural glass mineral, found its way from ash into groundwater and then into the trees. Thanks to silica, these dead trees live on in the fossil record.

The Great Plains are probably best known for their seas of grass, but astute observers will be able to sense mountains gaining height and volcanoes lighting up the sky. That story is neatly depicted in the bands of blue—and the petrified logs—that decorate these grassy plains.

Never take or move petrified wood or other fossils you might find in national parks (or on other lands, for that matter). It's a jerk move and against the law.

**Blue bands of bentonite are common in the park, especially along the Little Missouri River. A visit to the Petrified Forest Plateau to see petrified wood requires a short dirt-road drive followed by a hike of about a mile. Clay can make road conditions challenging. Always check with park staff about the status of unpaved roads.**

# BOTANY

## FIRST COMES THE FIRE

**WILDFIRE HAS A REPUTATION FOR** destruction and devastation, and for a legitimate reason—major fires can wreak havoc on key aspects of our lives, including destroying our homes. In many areas in North America, though, fire is actually essential for maintaining a healthy and diverse ecosystem. While you're exploring THRO (and other plains parks), you might see some recently burned areas. The charred landscape may appear lifeless and destroyed, but this postfire period is just a new beginning of a cycle that's been going on here long before humans arrived.

Wildfires occur one or two times per year in THRO. Often, they're caused by lightning strikes during intense Great Plains thunderstorms (see page 224). You might be imagining a vast field engulfed in enormous flames, but these fires have ebbs and flows. They can be difficult to find in their beginning stages, giving only the smell of smoke as evidence until the flames take over. Sometimes, controlled burns will be performed by the NPS with the help of other agencies, like the US Forest Service, in areas where fires haven't happened recently and where it's deemed necessary.

But why *is* fire necessary, anyway? Naturally occurring fires help maintain the plant diversity of the mixed-grass prairie, which benefits the wildlife that live here. Without fire, shrubs and trees can overtake the grasses, destroying food sources and habitat for wildlife. Fire suppression policies of the twentieth century—intended to protect people, their homes, and the land—eventually demonstrated that fire is crucial for many ecosystems. Wildlife and habitat concerns aside, when fires were suppressed, unburned areas accumulated so much fuel (plants and other biomass) that when fires did take hold they were incredibly intense and hot, causing more damage than regularly occurring fires would. Since changing those fire suppression policies, parks have put fire management policies in place to ensure the ecosystem is maintained.

## GETTING GRASSY

You might think all grasses are the same, but look closely and you'll see how diverse they are. Here are three common postfire grasses you might see in THRO.

**Little bluestem (*Schizachyrium scoparium*)** Grows in bunches and is 24 to 40 inches tall. When the plant is mature, its leaves turn a reddish brown and its seed heads are fluffy and white.

**Needle-and-thread grass (*Hesperostipa comata*)** Grows from a narrow stalk and has feathery strands at the top. These strands connect to a thicker portion of the grass, closer to the stalk, that makes it look like a needle and thread.

**Saltgrass (*Distichlis spicata*)** Usually grows in dense stands, generally sprouting up to 12 inches tall. Its leaf blades are sharp and crowded, growing up the entire length of its culm, or stem.

After a fire, grasses are the first to reappear. While it may look like nothing could have survived the flames, many grasses are simply "top-killed" by fire; their root systems remain intact. Those systems send up new growth after a fire. Needle-and-thread grass (*Hesperostipa comata*) is one such grass. Even if a full growing season passes after a fire, the grass still has a chance to recover in the second year.

Shrubs and trees appear after the grasses take hold. Often, when woodier plants are top-killed, they don't have the strong, established root systems of their grassy counterparts, so they rely on seeds to return. Rocky Mountain juniper (*Juniperus scopulorum*; see page 134) is a great example of this—it relies on animal transportation of its seeds to recover. Prescribed fires have actually been used in the park to prevent Rocky Mountain juniper from encroaching on (and ultimately taking over) grassland areas.

If you see or smell smoke in the park during your visit, remember that it's part of a healthy plains ecosystem, and that fires have been happening for millennia. Be sure to talk to park staff about smoke and fire that you see and how it will affect that area of the park.

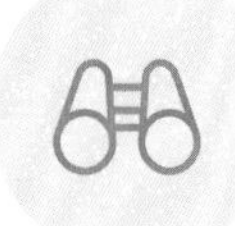

**Mixed-grass prairie is abundant in the park, so get up close to have a look—there are usually several species in one area. As you explore tall grassy areas, be sure to wear long pants tucked into your socks, and check for ticks afterward.**

# wind cave

STATE
SOUTH DAKOTA

ANNUAL VISITATION
656,000

YEAR ESTABLISHED
1903

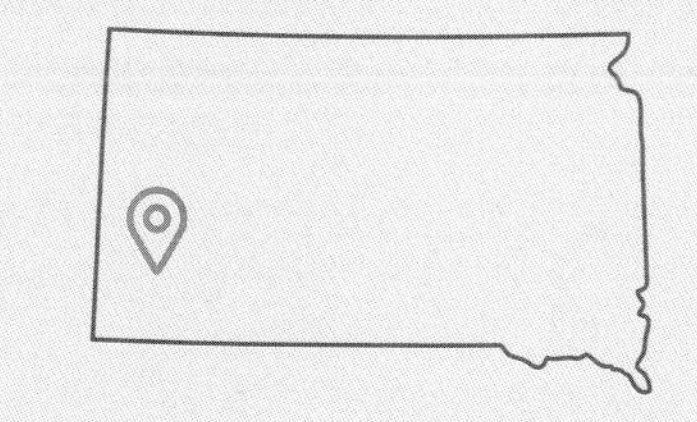

## a wild prairie atop a complex cave

| | |
|---|---|
| SUPERLATIVE | Wind Cave was the first cave to be designated a national park |
| CROWD-PLEASING HIKES | Elk Mountain Loop (easy); Lookout Point (moderate); Highland Creek Trail (strenuous) |
| NOTABLE ANIMALS | Black-tailed prairie dog (*Cynomys ludovicianus*); bison (*Bison bison*); burrowing owl (*Athene cunicularia*) |
| COMMON PLANTS | Ponderosa pine (*Pinus ponderosa*; see page 88); purple coneflower (*Echinacea purpurea*); sego lily (*Calochortus gunnisonii*) |
| ICONIC EXPERIENCE | Taking a ranger-guided tour in the cave |
| IT'S WORTH NOTING | The bison herd in the park is one of only four genetically pure herds in the world, meaning that they don't have domestic cattle in their genetic makeup. |

**WHAT IS WIND CAVE?** A complex maze of passageways, Wind Cave was created by water pooling in cracks inside rocks made from ancient ocean sediments. To date, almost 150 miles have been mapped, all under a small 1.25-square-mile patch of prairie. Explore the cave on a ranger-led tour to see unique formations, like calcite boxwork, and experience its namesake wind. WICA is more than the cave itself, though. Situated at the intersection of ponderosa pine forests and mixed-grass prairie, the park provides great opportunities to see wildlife such as bison, mule deer, and pronghorn. Aboveground or below the surface, WICA has something for everyone.

# WILDCARD

## THUNDERING ACROSS THE PLAINS

**THE GREAT PLAINS REGION IS** known for thunderstorms. A vast majority of the rain that falls here comes from fast-moving summer storms that frequently produce heavy rain and hail and streak the sky with lightning. The Plains' flat landscape makes storm fronts all the more dramatic since you can stand under blue sky and watch a wall of dark clouds approaching from nearly a hundred miles away. You can feel the wind pick up and the temperature change as the storm approaches. But why are thunderstorms so common at WICA and other places on the Plains?

To find the answer, you have to look far beyond WICA or even the Great Plains. The shape and topography of the North American continent has created something of a storm lab over the heart of the United States. The peaks of the Rockies act like a flood wall, forcing cool air from the Canadian Plains southward, where it clashes with warm, moist air rushing northward from the Gulf of Mexico. As warm and cool air meet, warm air is forced upward, where it cools and releases water vapor. The cooled air then sinks back down to be warmed again. This temperature-dependent circulation of air and water is called a *convection cell*. On a small scale, it's what forms clouds. On a much larger scale, like over the Plains, this convection forms and feeds thunderstorms.

Lightning—you know, the thing that produces thunder—is a key part of these storms. It is essentially a giant spark of electricity flowing between two differently charged surfaces. It's the same phenomenon as the small shock your finger produces when touching a metal doorknob after you shuffle across a carpeted room. In the case of lightning, the energy released is much, much more significant, and the surfaces involved are much bigger: a cloud and the ground, for example, or even two clouds. We still don't fully understand exactly how those charges build.

Normally, air is an electrical insulator, but lightning breaks it down and makes it a conductor instead. The characteristic zig-zag shape of lightning results from sections of air breaking down in sequence. It usually happens in discrete steps, each about 160 feet long.

## SENSIBLE STORM SAFETY

The first choice for safely viewing a thunderstorm is a fully enclosed space such as a hard-topped car or a building. (Tents and pavilions do not count.) If you find yourself out in the open, here are some tips for staying safe.

**1. GET LOW!** Find a valley or just a depression in the ground and get down there. Lightning is attracted to tall objects, so make yourself as short as possible.

**2. STAY OPEN!** Avoid being near trees or other tall objects. Lightning is most likely to strike objects that are tall and isolated.

**3. GET SPACE!** Make sure you are at least 100 feet away from other people. Ditto for any metal gear you're carrying, like hiking poles or metal-framed packs; these won't attract lightning, but they will conduct it, making the potential for injury much worse. Staying apart ensures that if one person is struck, others will be able to seek help for them.

**4. ASSUME THE POSITION!** Crouch on the ground with your weight on the balls of your feet. Tuck your head toward your chest and cover your ears. It's tempting to lie flat on the ground, but don't do it—electrical currents strong enough to cause serious injuries can travel through the ground after lightning strikes.

The thunder you hear after lightning is the sound of air around the lightning bolt rapidly and explosively expanding in response to the intense heat of the strike. Lightning strikes can reach temperatures five times hotter than the surface of the sun. Thunder occurs simultaneously with the lightning, but its roar seems delayed to us because sound travels so much more slowly than light.

Thunderstorms aren't just about what you can see and hear; they are also about what you can smell. Take a deep sniff right before the storm and you'll likely smell something sweet with a little bit of zip—that's ozone. It's a gas being swooshed down to the ground from the upper atmosphere by the movement of warm and cold air in the clouds. Have a sniff while it's actively raining and you'll smell petrichor—the distinct smell of raindrops disturbing scent molecules on the ground. It's not one molecule or compound, but the potpourri of a rainstorm. In WICA and other parks, you're likely to smell petrichor dominated by animal and plant molecules; in a city, asphalt, concrete, and metal tend to be the most recognizable scents. That damp, earthy smell after the storm passes? It's an organic compound called *geosmin*, produced by bacteria and blue-green algae in the soil; it's also what gives some veggies, like beets, their characteristic earthy flavor.

Watch a thunderstorm roll in across the prairie, breathe in the smells, feel the vibrations of thunder in your chest, and watch lightning light up the sky—WICA is storm country.

**Thunderstorms are common in the summer months but can happen any time of year.**

## WILDLIFE

### PRAIRIE DOGS—MORE THAN PESKY PESTS

Black-tailed prairie dogs can also be found in BADL.

**HUMANS AREN'T GENERALLY KNOWN FOR** our love of rodents. Mice, rats, and squirrels are often considered problematic pests and carriers of disease. The most recognizable rodents in WICA have had to win over humans' hearts to continue to thrive in the park. They're called black-tailed prairie dogs (*Cynomys ludovicianus*), and they are most certainly more than aggravating pests.

There are five species of prairie dog in North America. WICA is home to the black-tailed variety. Look at the ground in the park to know the color of their fur—they almost perfectly match the dirt in which they build their homes, or burrows. Those burrows are part of a complex underground system called a *town*. Within a prairie-dog town, there can be many neighborhoods, each home to one group of animals. There is still a lot to learn about how these burrows work, but we do know that prairie dogs use mounds at the openings to help prevent flooding and as a high lookout to scan for predators. They also build the openings strategically to take advantage of the wind of the plains for a natural ventilation system. No HVAC? No problem!

When you visit a prairie-dog town, you might hear some of their distinctive barks (hence the name *dog*). They're social animals and use different calls to alert each other about things such as predators—they may even have special calls for certain types of predators. Of course, they can't always avoid predators; prairie dogs as prey are incredibly important for sustaining wildlife populations in the park. Coyotes, birds of prey, and black-footed ferrets are just some of the creatures who count on prairie dogs as a big part of their diets.

Black-tailed prairie dog (*Cynomys ludovicianus*)

Black-footed ferrets were once believed to be extinct, but since a population was discovered in Wyoming in 1981, reintroduction efforts have been happening throughout the western United States. These predators rely heavily on prairie dogs for survival.

Unfortunately, during the twentieth century, prairie-dog habitat across the United States was reduced, largely because of farming and urbanization. Not only that, but prairie dogs have been the victims of a misconception that they compete with bison and cattle for grasses. In fact, black-tailed prairie dogs are integral to the entire grassland ecosystem; not only are they prey for various predators but they also improve the quality of the soil and have a positive impact on plant diversity in the area. These p-dogs are also threatened by the plague. Sylvatic plague is carried by fleas, and when it infects any species of prairie dog, it can wipe out more than 90 percent of the population, which can also be devastating for black-footed ferrets.

The good news is, the more you know about these special rodents, the more you can share that knowledge with others. Black-tailed prairie dogs are iconic animals of the prairie and worth more than a passing glance. And when you see them, imagine their elaborate underground tunnels and contemplate their social systems. For your safety and theirs, make sure to keep away from their mounds, as foot traffic can damage them—and rattlesnakes sometimes make a home there.

**Visit a prairie-dog town! There's one 1.2 miles north of the visitor center at the junction of US Highway 385 and SD 87, where you can pull off the road to observe prairie dogs in action. You can also ask park rangers where other towns are located. Be sure to stay on trails and avoid walking on the towns, which can cause irreparable damage.**

STATE
MINNESOTA

ANNUAL VISITATION
240,000

YEAR ESTABLISHED
1975

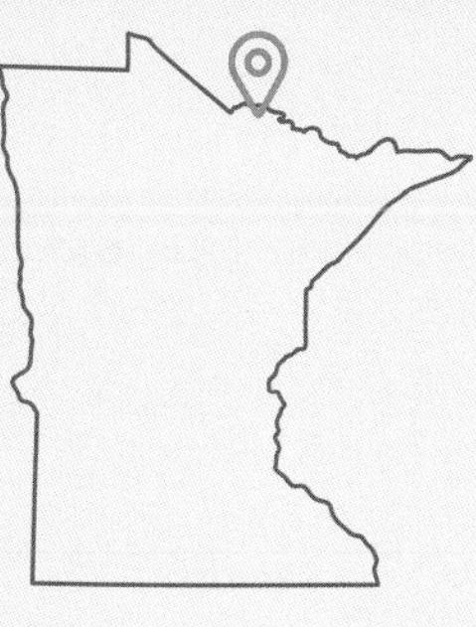

# Voyageurs

## a watery world of ancient geology

| | |
|---|---|
| CROWD-PLEASING HIKES | Echo Bay Trail (easy); Blind Ash Bay Trail (moderate); Cruiser Lake Trail (strenuous) |
| NOTABLE ANIMALS | Common loon (*Gavia immer*); moose (*Alces alces*); beaver (*Castor canadensis*) |
| COMMON PLANTS | White pine (*Pinus strobus*); inky cap mushroom (*Coprinus atramentaria*); blueberry (*Vaccinium angustifolium*) |
| ICONIC EXPERIENCE | Exploring the park on its famed waterways |
| IT'S WORTH NOTING | Winter brings extreme cold, but also a very good chance of seeing the aurora borealis (northern lights) light up the night sky. |

**WHAT ARE VOYAGEURS?** Voyageurs were seventeenth- and eighteenth-century fur-company employees who traveled this area by canoe. VOYA is named for them, but the park is so much more. This place is a water world, the flooded-out core of an ancient volcanic mountain chain. It is a maze of lakes, rivers, and streams best experienced from the water. Humans navigate the park by canoe, kayak, or boat during the warmer months, and skis and snowshoes in the frigid winters. Nestled along the northern US border, the forests on VOYA's islands and peninsulas are where the oaks and maples of southern forests give way to the spruces and firs that dominate colder climates. Pack your paddle and your parka—VOYA is a wet and wild place.

# BOTANY

## ALL CATTAILS ARE NOT CREATED EQUAL

**PEOPLE COME TO VOYA TO** be on the water, perhaps without realizing that the intersection of land and water is where you can find so much of this region's abundance. Many animals live in or reproduce near the park's wetlands. Some eat the wetland plants, while others eat the animals who eat the plants.

But all is not well in this biodiversity kingdom; an insidious invader has taken over. Narrowleaf cattails (*Typha angustifolia*) first showed up on the East Coast in the early 1800s and spread west through artificial waterways and human transportation. By the 1930s or '40s, the invading cattails started to appear on shorelines throughout VOYA. In the park (as elsewhere in the Midwest), narrowleaves quickly interbred with the native broadleaf cattails (*Typha latifolia*), forming a hybrid of almost supernatural strength. Hybrids (designated as *Typha x glauca*) grow taller, denser, and faster than natives and can thrive in a much wider range of water depths and conditions than either parent species.

The hybrid cattails spread swiftly and unstoppably through VOYA's wetland communities. Because invasive and hybrid cattails are nearly impossible to visually distinguish from the native species, the park conducted genetic studies in the early 2000s. By then, virtually every cattail in the park was a hybrid. The problem was not that the invading hybrids crowded out native cattails, but that they crowded out *everything*. Prior to the arrival of narrowleaves and hybrids, VOYA's broadleaf cattails grew sparsely in the wetlands, interspersed with a diverse array of wild rice, grasses, sedges, and rushes. But the invaders took over, leaving no room for anyone else. Instead of integrating into wetlands, they *became* the wetlands.

When you first see a stand of invasive hybrids, it probably won't register as anything other than a tight clump of wetland plants. But look closely: those are all the same plant. (Sometimes literally. Hybrids reproduce mostly by cloning, meaning that, sometimes, wetlands several acres in size will contain only one genetically identical plant.) The density of the stands prevents wildlife, like muskrats and waterfowl, from accessing the water effectively, forcing them to go elsewhere. For species that live under the water, like the northern pike, dense cattail stands reduce areas in which to spawn or hide from predators. When clumps of cattails detach from shore, they create huge floating mats that can damage park infrastructure and impair water traffic.

## BEFORE AND BEYOND THE CATTAILS

A healthy wetland hosts a few cattails, but many other plants as well. Look for these three native plants that have always called VOYA's wetlands home.

**Softstem bulrush (*Schoenoplectus tabernaemontani*)** Dense clumps of this sedge grow in water up to 3 feet deep. The spongy stalk (give it a gentle squeeze!) is generally not surrounded by leaves.

**Northern blue flag iris (*Iris versicolor*)** These showy flowers blossom in many shades of purple but are always decorated with yellow blotches. Look for pollinators, like bees, who are regular visitors to the blooms.

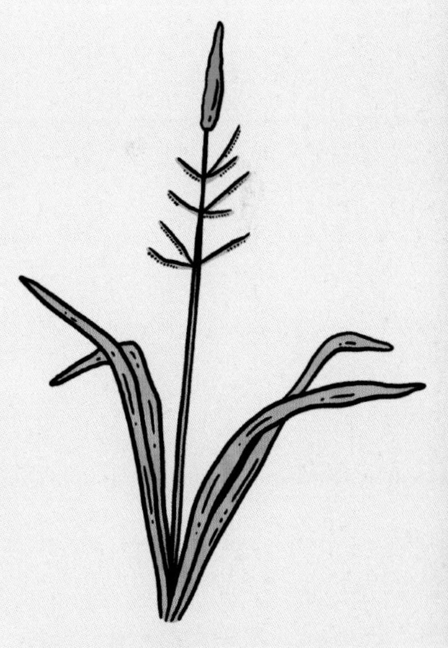

**Wild rice (*Zizania palustris*)** Wild rice has always been a vitally important wetland plant here, especially to the Anishinaabe (Ojibwe), for whom it is a dietary staple. The first human inhabitants in what is now VOYA, the Anishinaabe continue to gather wild rice from these wetlands today.

Things are changing, though. In 2016, the park began an ambitious removal and restoration program. Hybrid cattails are being chopped, flooded, or burned to a toasty crisp. (Sometimes controlled fire actually is the answer!) Usually a combination of methods has to be employed to ensure that the plants cannot regenerate.

Plant-killing herbicides aren't an option because of the protected status of VOYA's waters. Dams outside the park also complicate efforts to flood and drown the hybrids.

If invaders can be cleared from wetlands, something magical happens: native plants reclaim the area from seeds that have laid dormant on the lake bottom for decades. In places where regeneration is trickier, park employees help jumpstart the process with native seeds and plants collected from other parts of the park or other nearby areas. Finding local seeds is crucial—they stand the best chance in northern Minnesota's, ahem, chilly climate.

Less than a hundred years after the invasive's arrival, hybrid cattails have significantly altered the shorelines of VOYA. Restoring balance to the wetlands here in the form of plant diversity is a complicated process, but it's well under way. Soon VOYA's watery margins will once again host the plants and animals that have depended on them since the last retreat of the glaciers thousands of years ago.

**A little less than half of VOYA is water, so wetlands abound—including a large one right outside the Rainy Lake Visitor Center. That's also a good place to check for ranger programs and get updates about the cattails.**

STATE
MICHIGAN

ANNUAL VISITATION
26,000

YEAR ESTABLISHED
1940

# isle royale

## an island of volcanic rock scarred by glaciers

| | |
|---|---|
| SUPERLATIVE | Isle Royale is the largest island in Lake Superior |
| CROWD-PLEASING HIKES | Windigo Nature Trail (easy); Suzy's Cave Trail (moderate); Greenstone Ridge Trail (strenuous if you do it all) |
| NOTABLE ANIMALS | Moose (*Alces alces*); painted turtle (*Chrysemys picta*); sandhill crane (*Antigone canadensis*) |
| COMMON PLANTS | Common butterwort (*Pinguicula vulgaris*); prickly saxifrage (*Saxifraga tricuspidata*); balsam fir (*Abies balsamea*) |
| ICONIC EXPERIENCE | Being on the water, either by ferry, boat, canoe, or kayak |
| IT'S WORTH NOTING | The park has a lot of shipwrecks that scuba divers can explore. |

**WHAT IS ISLE ROYALE?** Isle Royale is the place to go if you're looking for solitude in the lower 48. This isolated island park on Lake Superior is only open for about seven months of the year due to harsh winter conditions. When it is open, it's accessible only by boat or plane, which means you won't be bringing your car along for the trip. The product of ancient volcanic eruptions later carved by multiple glaciations, the park's rugged terrain and abundant water delight backpackers and paddlers alike. At night, ISRO benefits from the lack of developed areas and has some of the darkest—and starriest, when it's clear—skies in the country.

# BOTANY

## WHAT THE GLACIERS LEFT BEHIND

If you're really into glaciers, see the pieces on glacial erratics in YOSE (page 35), and glacial lakes in SEKI (page 62)!

**A BILLION OR SO YEARS** ago, basaltic lava flows formed what is now ISRO. Of course, a lot of geologic happenings have taken place since then, but many of them are still a mystery because of at least four major glaciations over the last 2 million years. Each advance and retreat of the ice reshaped the land and erased or obscured what had come before. Most of what you see today is the result of the most recent glaciation of the area. Known as the Wisconsin Glaciation, it happened about 11,000 years ago, and you can see its impact on the land.

If you dip your toe into Lake Superior, you'll notice that it's pretty chilly no matter the time of year. The lake's cold waters create a cool microclimate along the rugged Superior shoreline, where rocks are a more common sight than fertile soils. Surprisingly, there are some plants that prefer these harsh, chilly areas. These plants, known as *arctic disjuncts,* are usually found in arctic environments much farther north. There are various hypotheses about why they're here, but no one knows for sure. It's possible these species lived south of the most recent ice sheet, and when that retreated, some remained in this region while others moved north to the Arctic. Another possibility is that the species moved south from the Arctic to this region after the glacial retreat. Most likely it is some combination of these hypotheses, depending on the plant species; regardless, it's fascinating to have these arctic experts living so far south.

Around twenty species in the park fit this arctic disjunct bill. Not surprisingly, their growing season is really short (and their flowering seasons are even shorter—usually just a couple of weeks), which means that seeing them is a rare treat. The prickly saxifrage (*Saxifraga tricuspidata*) is a flowering beauty that occurs only in ISRO in the lower 48. It's also called three-toothed saxifrage, because its smooth green leaves (1 to 2 cm long) have three sharp spiny "teeth" on their ends. Look for stunning flowers in late May through the end of June—they're white with five petals with lovely red or purple dots.

ISRO also has fruiting plants, including the black crowberry (*Empetrum nigrum*). This is a low-lying evergreen shrub that forms mats and has leaves that are tiny and tubular in shape. In July and August, the plant grows fruits that are berry-like, round, and black, about the size of a blueberry. In fact, they're members of the same family as the blueberry. Don't eat the fruit, though; it's not tasty!

Of course, no list of unusual plants is complete without a carnivore in the mix. The common butterwort (*Pinguicula vulgaris*) gets its nutrients from unfortunate insects that find themselves trapped on its sticky leaves, which lie flat on the ground. The stickiness comes from enzyme-secreting glands that capture the insects and then break them down for digestion. There's more to the plant than these useful leaves, though; in June and July, they grow a solitary purple flower with a white spot at the mouth atop a leafless stalk that may reach 4½ inches tall.

Common butterwort with dead insects ready to be digested.

While you're hunched over admiring these short-lived plants up close, remember that they're a window into an ancient world, inviting you to consider ISRO's icy past.

**During the growing season, look for these plants along the rocky shores of the park.**

## GREENSTONE: SUPERIOR VOLCANIC GEMS

ISRO AND THE nearby Keweenaw Peninsula are two parts of the same lava flow. The Greenstone Ridge Lava Flow, up to 800 miles thick, is the backbone of ISRO, travels underneath Lake Superior, and pops up 50 miles later at the Keweenaw Peninsula. Trapped within parts of this basaltic flow are amygdules, or tiny gas bubbles filled with other materials. Some of the amygdules are filled with the green mineral chlorastrolite, better known as Isle Royale greenstone. This beautiful and rare stone, found only in ISRO and the Keweenaw Peninsula, is Michigan's state gemstone. If you happen to spot it in the form of small pebbles along ISRO's shore, leave it where it is to preserve it for future generations—also, removing anything from the park is illegal and could result in a fine.

# great smoky mountains

**STATE**
**TENNESSEE & NORTH CAROLINA**

**ANNUAL VISITATION**
**11.4 MILLION**

**YEAR ESTABLISHED**
**1934**

## a temperate rain forest on ancient mountains

| | |
|---|---|
| SUPERLATIVE | GRSM is the most visited national park (by a landslide!) |
| CROWD-PLEASING HIKES | Andrews Bald (easy); Spruce Flat Falls (moderate); Charlies Bunion via the Appalachian Trail (strenuous) |
| NOTABLE ANIMALS | Black bear (*Ursus americanus*); eastern screech owl (*Megascops asio*); thirty-plus species of salamander |
| COMMON PLANTS | Rosebay rhododendron (*Rhododendron maximum*); flame azalea (*Rhododendron calendulaceum*); mountain laurel (*Kalmia latifolia*) |
| ICONIC EXPERIENCE | Taking in the views from Clingman's Dome Observation Tower |
| IT'S WORTH NOTING | In 1983, the park was one of the first places to be designated a UNESCO World Heritage Site. |

**WHAT ARE THE GREAT SMOKY MOUNTAINS?**
The Smokies are one part of the Appalachian Mountains. They were once as high as the Rockies, but in the course of 300 million years the Appalachians have been eroded to rounded humpbacks. In the southeast United States, rain-fed forests cover the low mountains and give off the Smokies' namesake haze. The temperate rain forests of GRSM are a biodiversity hotspot; driving from the park's lowest elevations to its summits is the equivalent of driving from Georgia to Maine in terms of diversity of plant and animal life. Some spots in the park get more than 80 inches of precipitation a year, so don't forget to pack your raincoat along with your curiosity when you head out to explore!

# BOTANY

## TOWERING TREES

**WALK EVEN 20 FEET ALONG** a forested path into the Smokies and you can be enveloped completely by green. To the untrained eye, most forests look pretty similar. But here the forests are special, offering a fleeting glimpse of the majestic trees that blanketed eastern North America before colonization. About a fifth of the park area escaped the logging trucks, plows, urban development, and cattle grazing that wiped out huge tracts of trees in the eastern United States. Though there isn't one definition for what *old growth* means, the term is used to describe forests such as these that have not been disturbed by Euro-Americans.

Left to their own devices, old-growth forests have flourished in GRSM's rough, rain-soaked terrain. Abundant rainfall, moderate temperatures, rich soils (look at all that leaf litter!), and minimal human interference have all helped the trees here reach astonishing sizes. The trees in GRSM regularly make the list of the largest trees in the United States, compiled by the American Forests group. Organized by species, the list ranks trees based on a combination of height, circumference, and crown spread. The trees' specific locations aren't shared, in order to protect them, but past listees include a yellow buckeye with a trunk almost 20 feet around and a black cherry more than 132 feet tall. Some especially prime spots in the park, like the Cataloochee Valley, have trees of various species that *average* more than 150 feet tall—that's almost half as tall as the Statue of Liberty.

Many old-growth areas in the park were spared European intrusion because they were difficult to reach or contained trees that were considered "not valuable" by the logging companies. (As if a tree could be a waste of space!) Today, reaching most old-growth groves still involves at least a short hike. Along hiking trails, no sign will alert you to old-growth areas, but you will quickly learn to slow down and spot these beautiful old forests.

Groves that were spared because they didn't contain the "right" kind of timber are now home to some of the most ancient trees left in the eastern United States. Even trees that aren't especially large can be easily 300 years old.

## GREAT SMOKY TREES

Leaf shape is the easiest place to start identifying a tree. For very tall trees, look around on the forest floor for fallen leaves. Always remember to stay on designated trails and never pick leaves or break off twigs.

**Tulip tree (aka yellow poplar; *Liriodendron tulipifera*)** Look for their tall, straight, almost perfectly round trunks with branches that start high off the ground. Look closely and you may see what looks like white or silver chalk dust in the grooves of the bark.

**Black cherry (*Prunus serotina*)** Mature trees grow very tall and have a distinctive gray-black flaky bark—it looks like large cornflakes are glued onto it. On small branches and younger trees, look for smooth bark with fine white lines.

**Yellow buckeye (*Aesculus flava*)** Look for leaves that resemble fingers on a hand—five leaflets radiating out from a single stem. Mature trees often have bark that is broken into plates and scales.

**Mountain or Carolina silverbell (*Halesia carolina* or *Halesia monticola*)** Silverbells in the southern Appalachians grow so much larger than in other places that they were once considered a separate species. Look for dark reddish brown scales on the bark of mature trees.

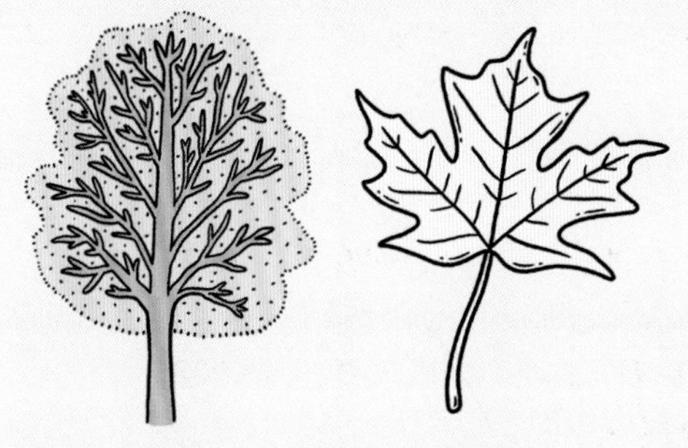

**Sugar maple (*Acer saccharum*)** Look for large trees with blackened bases and bark with curled edges. Younger trees have smooth gray bark. The Sugarlands area of GRSM is named for a grove of these trees that were formerly tapped for their sugary sap.

**Northern red oak (*Quercus rubra*)** These oaks can be distinguished from the other oaks here by their long, wide, slightly concave plates of bark. Stand underneath and look up to see silver streaks along the bark plates and big branches.

Watch out for poison ivy as you explore! It's common in the park, in both old and young forests. Never touch a vine or shrub you don't recognize.

Really, really big trees are an obvious clue, but the hallmark of an old-growth forest is diversity—lots of species of trees in lots of different life phases. Look for young trees, dead and dying standing trees, mature stately trees, and lots of leaf litter, logs, and dead branches on the ground. Look up. Even among a grove of huge trees, the canopy of an old-growth area will still have a somewhat open feel. Where a large tree is decaying on the ground, see if you can still make out the gap it left in the canopy when it fell. When a giant falls, the light that is now able to pour in jumpstarts a race for resources between plants and seedlings that may have been dormant for many years on the forest floor.

Towering trees notwithstanding, it's important to recognize that even old-growth forests have felt the impact of humanity. Invasive species have crept into this landscape, sometimes with disastrous consequences. Such was the case with the hemlock woolly adelgid (*Adelges tsugae*), an insect accidentally introduced from Asia that has devastated the park's renowned hemlocks. Depressing stands of dead hemlock trees are not difficult to spot in GRSM.

The good news is that there is plenty of old-growth forest in the park to explore. Stay on the trail, leave no trace, and, for goodness' sake, don't take anything from the forest (we're looking at you, leaf collectors). Savor your time among trees with deep roots and even deeper history; this is a forest at full strength.

**Some of the best hikes to see old growth are Cove Hardwood Nature Trail, Ramsey Cascades Trail, Albright Grove Loop Trail, Boogerman Trail, Caldwell Fork, and Grotto Falls. Check in at a visitor center about trails and drives suitable for you.**

## WILDLIFE

### DANCING IN THE DARK

Most of the United States east of Kansas have a number of firefly species. In Western states, they're much less abundant.

**BIOLUMINESCENCE, OR THE BIOCHEMICAL EMISSION** of light by an organism, is one of the coolest processes of our natural world. It happens all over the planet—in the ocean, on land, and in the air. If you stay up past sunset during your visit in GRSM, you're likely to enjoy the bioluminescent light shows put on by the nineteen species of fireflies, or "lightning bugs," in the park. Whatever you call them, fireflies are one of the most famous types of beetle in the world.

How *do* these little beetles make their light? Put very simply, they initiate a super-efficient chemical reaction in their abdomen (in a spot called their *lantern*) that emits light. This reaction mainly involves luciferin (a molecule), luciferase (an enzyme), ATP (for energy transfer), oxygen, and nitric oxide. Think of it as teeny tiny lightning flashes that happen in those bug butts, only their light doesn't release heat. It's some of the most efficient light on the planet.

These beetles (and their larvae) presumably use bioluminescence for two reasons, but no one knows for sure. One reason is to let predators know they taste bad, and the other is to attract a mate. In many species, male fireflies perform the classic fly 'n' flash to let females know they're interested. The ladies flash back, the male lands, and they mate.

The flagship firefly festivities in GRSM are the synchronous firefly (*Photinus carolinus*) displays in June. As the name suggests, the males flash in unison—six bright flashes followed by six seconds of "silence" repeated over and over again. Science still hasn't explained exactly why the males flash in unison, but anyone who's seen it can agree that it's a magical spectacle.

The blue ghost firefly (*Phausis reticulata*) takes the cake for being the eeriest lightning bug of the bunch. Look for their continuous blue glow around ankle height. They stay out later at night and fly for longer in the season than their synchronous counterparts. It's also worth noting that once female blue ghosts lay their eggs, they continue to nurture and protect them until they die, which is fairly uncommon in the insect world. Go, moms, go!

BLUE GHOST GLOW

[ 6 flashes ] [ 6 seconds ]

SYNCHRONOUS FLASH PATTERN

Two of the firefly flashes you might see in GRSM

As you're out enjoying the darkness of the park, you're also likely to see the big dipper firefly (*Photinus pyralis*) in June and July, and sometimes all the way into September. Big dippers are probably the most recognizable firefly in the United States, with a massive range from Texas to South Dakota all the way to Florida and New York. Look for the males' longer yellow flash around waist height.

If you live in the fireflies' eastern range, keep an eye out for them once you return home. With thousands of species and a short-lived season, the best way for scientists to learn more about them is by collecting vast amounts of data. Ordinary people are able to participate as citizen scientists to help observe fireflies in action—if you're interested, run a quick search for current citizen-science opportunities.

**Look for big dippers during the summer (peak in June and July) in open fields and campsites, starting about 90 minutes before sunset and as late as midnight.**

**Blue ghosts are especially prevalent at higher elevations during late spring and summer. They prefer leafy forest habitats. Brightest displays are about 30 minutes after sunset and last about 90 minutes.**

**Though synchronous fireflies can likely be seen elsewhere during the summer, the most accessible views are from the Elkmont area. Access is closely managed during peak displays (two weeks in June). Check the NPS's Firefly Event website for details.**

# GEOLOGY

## GO CHASING (SANDSTONE) WATERFALLS

**AROUND 800 MILLION YEARS AGO**—when life on Earth was just weird tube-shaped things floating around—an ocean stretched out from the eastern shore of North America. Somewhere offshore, sediments drifted to the ocean floor for a few hundred million years. Sometimes it was quiet; sometimes debris flows swept wood, chunks of rock, and hapless shelled critters into the depths. Eons later those sediments were hurled upward as Africa slammed into North America, part of a series of tectonic collisions that gave birth to Pangaea, the ancient mega-continent. The power of that event raised the Appalachians, including the Smokies, to heights that likely approached that of the modern Rockies. It also squeezed and cooked those ocean sediments until some transformed into a hard, blue-tinted rock called Thunderhead Sandstone. (It's actually a metasandstone, meaning that it's been heated and altered—not that it says a lot of deep things.)

Over time, those high mountains eroded into the rounded Appalachians of the present. Thunderhead Sandstone remains, though, forming many of the cliffs and ledges in the park and capping peaks such as Clingman's Dome and Mount Guyot. It looks a lot like other closely related sedimentary rocks in the park, but it's special: Most of the waterfalls in the park fall over Thunderhead Sandstone.

The erosion-resistant properties of Thunderhead Sandstone join forces with the Smokies' mountainous topography and ample rainfall to create a park that is absolutely dripping with waterfalls (pun intended). There are more than a hundred prominent falls in GRSM, and many more unnamed cascades. Much of the rain that falls in the higher reaches of the park is channeled into 2,000 miles of streams that wend their way down the mountains. And where those streams meet beds of Thunderhead Sandstone, you are very likely to find a waterfall.

Spend a minute standing in front of a waterfall. At virtually any falls in the park, the water runs along the contact point between Thunderhead Sandstone and whatever less resistant rock lies above it (often visible on either side of the upper part of the falls). Sidle up to a wall of Thunderhead Sandstone to get a good look at its famous blue-tinted quartz grains. Near waterfalls, look for white chunks of feldspar and darker pieces of slate embedded in Thunderhead Sandstone—rock fragments swept into those ancient ocean sediments. Drier outcroppings are good places to look for pebbles—some as big as ½ inch across—tucked neatly into the linear bands of the rock.

Thunderhead Sandstone isn't immune to erosion, because no rock is. Along with the rest of the rocks in the Smokies, it is being slowly and irreparably dismantled, grain by grain, and carted away in streams and rivers. In a process that's been going on for millions of years, approximately 2 inches' worth of the Smokies washes away every thousand years. The waterfalls and blue-tinted rock walls of Thunderhead Sandstone you can see today are just one small piece of the much larger story of how our planet's surface is constantly in flux. Take time to enjoy these rocks and waterfalls in this present moment—we know it won't last.

**Waterfalls are common in the park. Some especially noteworthy ones include Ramsey Cascades, Grotto Falls, Abrams Falls, and Laurel Falls. Hikes to waterfalls range from easy to strenuous. Some falls, such as the Sinks, can be viewed from a car.**

STATE
**MAINE**

ANNUAL VISITATION
**3.5 MILLION**

YEAR ESTABLISHED
**1919**

## rounded mountains & rugged shorelines

| | |
|---|---|
| SUPERLATIVE | ACAD was the first national park designated east of the Mississippi River |
| CROWD-PLEASING HIKES | Great Head (easy); The Bubbles (moderate); Beehive Trail (strenuous) |
| NOTABLE ANIMALS | Periwinkle (*Littorina* spp); black guillemot (*Cepphus grylle*); raccoon (*Procyon lotor*) |
| COMMON PLANTS | Rock polypody fern (*Polypodium virginianum*); paper birch tree (*Betula papyrifera*); lichen (*Cladonia* spp., among many others) |
| ICONIC EXPERIENCE | Catching the sunrise with the crowd on Cadillac Mountain or from a more secluded spot |
| IT'S WORTH NOTING | At low tide, you can take a short and beautiful walk across a sandbar to Bar Island. |

**WHAT IS ACADIA?** Situated mostly on Mount Desert Island, ACAD is the heart of the North Atlantic coast. Ocean temperatures here almost never get hotter than 60°F, and there is only one sand beach for your sunning pleasure. Rich waters feed populations of fish and sea mammals, while tide pools shelter their own fiercely competitive mini worlds. The park is famous for its rounded granite mountains and its miles of old carriage roads that still crisscross the island, a holdover from Acadia's stint as a robber barons' playground in the late nineteenth and early twentieth centuries.

# GEOLOGY

## THE SHATTER ZONE

**AROUND 420 MILLION YEARS AGO,** the land that would become Mount Desert Island was really going through the wringer. Several miles below the surface, a huge chamber of magma began to form and fill. Over time, more and more molten granite squeezed into the chamber, until the whole thing was filled to bursting. Overwhelming pressure fractured the rock above the chamber, and magma exploded outward, forming a caldera 10 miles across. With the pressure released, shattered rock on the top and sides collapsed into the magma chamber. It happened fairly quickly, probably in a matter of days or weeks. Cue the guitar solo because you can still see that magma chamber today. In fact, you can stand on the very edge of it.

It's called the *shatter zone*—as rock 'n' roll as geologic names get—and today it forms a semicircular arc around the southern and eastern sides of Mount Desert Island. Inside that ring is the granite core of the magma chamber, which has been uplifted and eroded into the rounded domes of Mount Desert Island. You read that right: Cadillac Mountain is a pile of cooled magma.

The shatter zone is the edge of that magma pile, where fractured blocks of rock from the sides and top of the caldera collapsed into molten granite. It ranges from a thousand feet thick to more than a mile deep in some places. This semicircular arc of granite is loaded with fragments of other rocks ranging from golf ball–size to blocks hundreds of feet across. Think of irregular chocolate chunks dropped into thick cookie batter and partially stirred.

These chunks are mostly exposed around the edges of the granite, in places where the magma cooled relatively quickly. Because it cooled quickly, the granite in the shatter zone looks a little different than the same rock elsewhere on the island. To find shatter-zone granite, keep an eye out for light gray, almost white granite that doesn't have large flecks of white, black, or translucent minerals.

# COBBLING TOGETHER THE SHATTER ZONE

The large beach rocks, known as cobbles, on Little Hunters Beach are broken-off pieces of the shatter zone and the various kinds of rocks present on the island. In other words, they are a snapshot of ACAD's violent volcanic past. Always use caution when exploring near water, and take care not to rearrange or greatly disturb the cobbles. Definitely never take them! While you're on the beach, listen for the rattling sound of big waves tumbling the cobbles underwater.

### Ellsworth Schist

Look for alternating bands of light and dark in chaotic swirls, infused with a deep-green hue. This is 500-million-year-old seafloor muck that has been heated, compressed, twisted, and finally eroded and tumbled down to this beach.

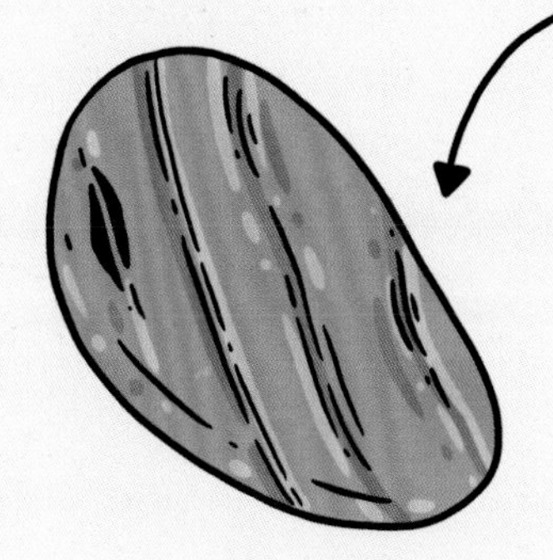

### Gabbro

Look for dark, somewhat shiny rocks, sometimes marked by light-colored streaks. Look closely to see individual black and gray mineral grains. Blocks of gabbro can be seen inside granite along the western side of Little Hunters Beach.

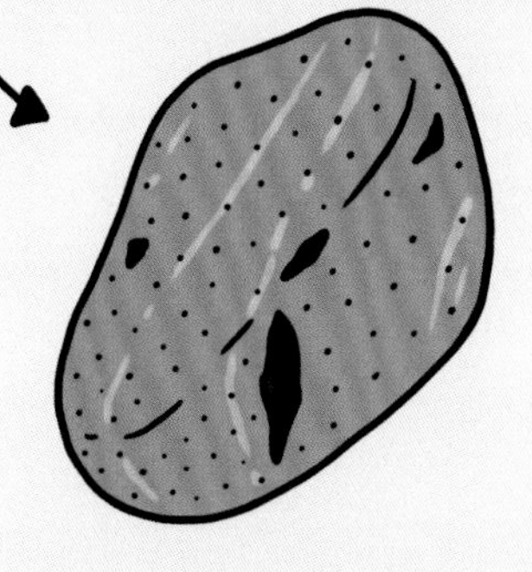

### Cadillac Granite

Look for pink to greenish gray rocks uniformly speckled with white, black, pink, clear, and gray. This granite made up most of the magma chamber and cooled relatively slowly, giving it time to form those multicolored mineral chunks.

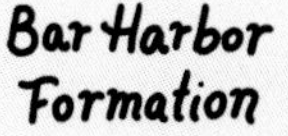


Lots of iron in these rocks means that they rust when exposed to air. Look closely to see tidy, distinct layers. Chunks of BHF rocks encased in granite are visible along the west side ledges of Little Hunters Beach.

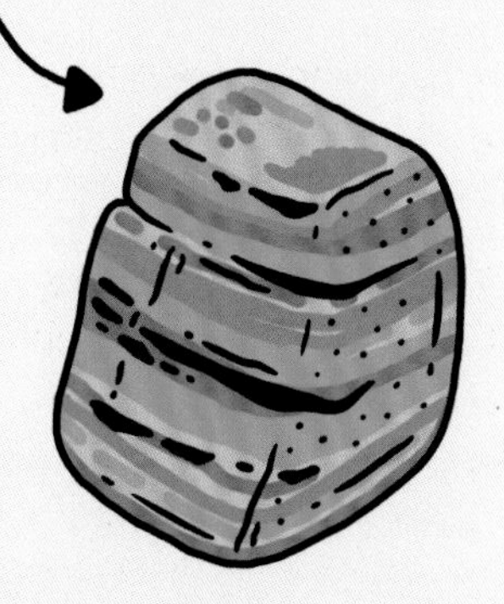

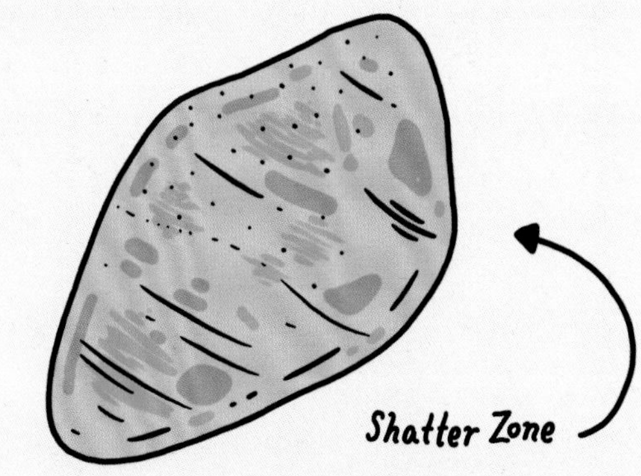

### Shatter Zone

Look for light-colored rocks with the look of granite. These rocks are usually more angular than other cobbles, since they come from nearby. Often rusty pieces of Bar Harbor Formation and/or dark pieces of gabbro are visible.

And, of course, look for chunks of other types of rocks within the granite! Geologists call them *inclusions*. Just as you might put more than one kind of chocolate into cookies, there are multiple kinds of rock locked inside the granite. Look for rusty red inclusions from the sedimentary Bar Harbor Formation. Darker gray masses are gabbro: igneous rock that arrived on the scene just before ACAD's granites. Some rock chunks began to melt in the hot magma, which rounded their edges. Other pieces have jagged edges and obviously weren't vibing with the magma in quite the same way. Many shatter-zone exposures look chaotic, with lots of bits and pieces swirled around.

ACAD offers plenty of idyllic views and seaside escapes, but now you know to listen carefully. Under the sound of surf meeting shore, it's possible to imagine the faint but unmistakable sound of an ancient volcano unleashing its fury.

**Little Hunters Beach has the best and most easily accessible views of the shatter zone along the ledges on the far western side (to the right as you're facing the water). Other places to see it include Otter Cliffs, the east end of Sand Beach, near the parking lot for the overlook at Schooner Head, and Western Point south of Blackwoods Campground. It is also well exposed on both sides of the entrance to Northeast Harbor; look for it on the boat ride out to the Cranberry Isles.**

## WILDLIFE

# CONSIDER THE SEA SNAILS

**IF YOU SPEND ANY TIME** tide pooling in ACAD (and you absolutely should), you'll have the potential to encounter any number of exciting species. Sea stars, crabs, anemones, barnacles, and green, brown, and red algae all call ACAD's tide pools home. As you walk around, you'll notice someone else lurking. Everywhere. SEA SNAILS! Marine snails are all over the place in ACAD, and there's more to them than meets the eye.

Since sea snails are so abundant and move so slowly, you can pretty much count on seeing some of these gastropods. Most of the snails in the park, including three species of periwinkle (common, smooth, and rough), have the typical coil-shaped shell. The common periwinkle (*Littorina littorea*) is an herbivore that scoots along and eats algae using its radula, or toothed tongue-like appendage. There are also tortoiseshell limpets (*Testudinalia testudinalis*), superabundant snails with shallow, conical shells that they use to scrape up algae to eat. They also use those shells to bulldoze barnacles in their path.

If you're more interested in the periwinkles' carnivorous counterparts, you can find evidence that they've been in the area. Look for empty shells with a tiny hole in them. Atlantic dogwhelk (*Nucella lapillus*) are carnivorous snails that use a mouth-like appendage to drill into the shells of other snails and suck them out. This might seem vicious, but snails are obviously not at the top of the food chain of the Atlantic coast. They do have some pretty neat tricks up their shells to avoid being tonight's dinner, though.

Common periwinkle shell (*Littorina littorea*)

Snails are more intelligent than you might think. They're able to take cues about local predators and adapt to them. For instance, snails can sense that invasive crabs are in the

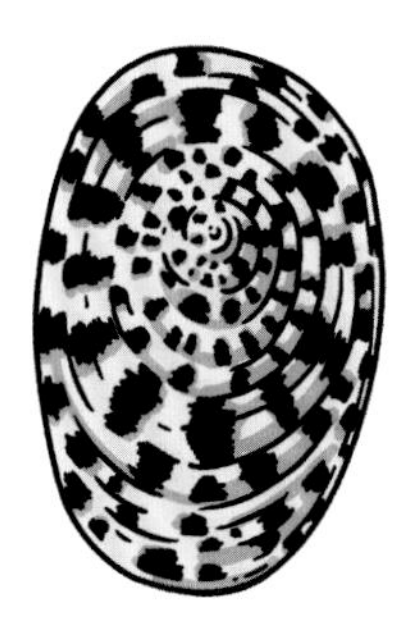

Tortoiseshell limpet shell (*Testudinalia testudinalis*)

area (from clues such as crab pee), and they're able to move to a safer place to forage before even interacting with a crab. They're also able to scoot away (slowly) and hide in cracks and crevices that predators can't reach. Some snails can even grow thicker shells within 90 days to protect themselves when they know predators are nearby. Given that snails live between 5 and 10 years, that's pretty quick!

Give snails more than a passing glance as you tide pool, and recall that these areas are their own unique habitats, sometimes treacherous and always exciting. If you're wondering why there are so many snails here, recall that it's illegal to hunt or gather anything within the park's boundaries, so humans are not permitted to collect snails for the eating. Before setting out on your tide-pooling quest, remember to take a look at a tide chart to find out when the tide will be low so that you can safely observe these watery worlds. Make sure you watch out for wildlife, like barnacles, and step carefully—the rocks are slippery!

**Sea snails are abundant in the tide pools in the park, which are easily accessible at Wonderland and Ship Harbor, and some can be found along the Bar Harbor Sand Bar at low tide.**

STATE
**VIRGINIA**

ANNUAL VISITATION
**1.3 MILLION**

YEAR ESTABLISHED
**1935**

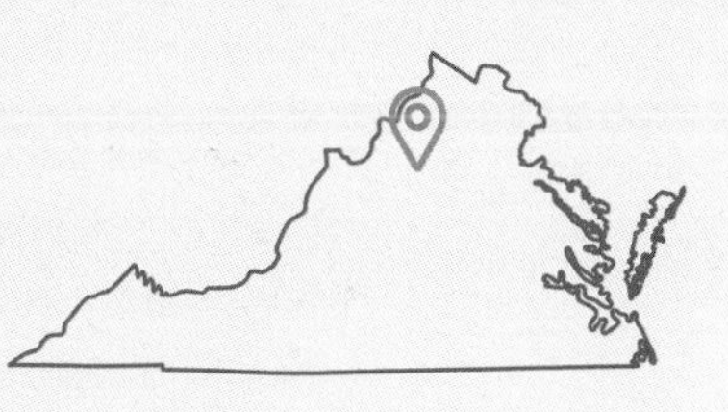

# Shenandoah

## forests & waterfalls atop ancient mountains

| | |
|---|---|
| CROWD-PLEASING HIKES | Frazier Discovery Trail (easy); Dark Hollow Falls (moderate); Riprap-Wildcat Ridge (strenuous) |
| NOTABLE ANIMALS | Black bear (*Ursus americanus*); downy woodpecker (*Picoides pubescens*); eastern bluebird (*Sialia sialis*) |
| COMMON PLANTS | Pink lady's slippers (*Cypripedium acaule*); black cohosh (*Actaea racemosa*); hairy lip fern (*Cheilanthes lanosa*) |
| ICONIC EXPERIENCE | Driving Skyline Drive and stopping at the overlooks |
| IT'S WORTH NOTING | You can hike a portion of the Appalachian Trail on your visit—the park includes more than 100 miles of the trail. |

**WHAT IS SHENANDOAH?** Shenandoah was originally a name associated with the Oneida Indian Nation in central New York, but the moniker was later transferred to a lush and rolling part of Virginia. SHEN protects a stunning stretch of the Appalachian Mountains in Virginia known as the Blue Ridge. Skyline Drive, the park's main road, follows ridges and valleys for over 100 miles, unrolling panoramas of the Shenandoah Valley to the west and the gentle hills of the Virginia Piedmont to the east. The park's rich forests are alive with songbirds, deer, and bears and are dotted with wildflowers and clear mountain waterfalls. Though SHEN seems isolated and idyllic, it's a mere 75 miles from Washington, DC. Even so, its night skies are some of the best in the East—be sure to look up after dark!

# WILDLIFE

## BEARS, BEARS, EVERYWHERE!

**BEARS ARE SOME OF THE** most widespread mammals in North America, and SHEN has one of the densest populations of black bears (*Ursus americanus*) in any national park. You may be intrigued by bears, intimidated, or indifferent, but it's impossible to ignore their presence. Look for bear evidence such as scat, but also keep your eyes open for the bears themselves—some research suggests there may be as many as one bear for every one square mile of the park (that's a lot of bears!).

Black bears used to thrive all across North America, numbering up to 2 million prior to European colonization. Shocking no one, after Europeans arrived, the population of black bears (and everyone else) plummeted due to hunting and habitat loss. By the early 1900s, within what would become SHEN's boundaries, there were no more black bears. A couple (literally two) brave bears wandered from the Allegheny Mountains (most likely) into the park after it was established in 1937. Since then, with management and conservation efforts, hundreds of black bears have been able to call this area home again. The park estimates there are 300 to 600 bears within its boundaries—of course, park populations fluctuate by year and season.

Black bears are crucial to a successful ecosystem. They thrive in the fertile forests of SHEN and live at the top of the food chain, eating animals such as groundhogs and other smaller mammals. They're opportunistic eaters, though, so they also eat insects, roots, berries, flowers, and carrion (the flesh of dead animals). By eating berries, they help with seed distribution, and they also help plant decomposition in the park by tearing apart decaying logs searching for insects and uprooting plants.

Since black bears are protected from hunting within the park, they're safer here than in other parts of Virginia (and the country). It is very possible that you'll see a bear at some point during your visit, which is exciting and exhilarating, but that comes with its own potential problems. Park rules

## BLACK BEARS AND YOU

Being prepared for a black-bear encounter is hugely important for your trip to SHEN, even if you plan to stay in developed areas. Keep at least 50 yards away from black bears that you see. Please also note that this guide applies only to black bears—if you're in a park with grizzly bears (such as GLAC), gather information from the visitor center or a ranger about what to do if you encounter them, since the two species are very different. Also be sure to check with every park about their bear-spray policies before you visit, and if you plan to bring bear spray, make sure you know how to use it.

Be very cautious if you see a mother bear with cubs, and do not get between them.

IF YOU ENCOUNTER A BLACK BEAR

- Make sure the bear knows you're human, not prey.
- Do not run.
- Talk to the bear in a normal voice.
- Do not attempt to climb up a tree, as black bears are great climbers.
- Give the bear a route to escape.

IF THE BEAR DOESN'T LEAVE

- Make yourself as large as possible.
- Make loud noises; bang things such as pots and pans if you have them.
- If the bear attacks, fight back, going for its eyes and nose—do not play dead.

regarding wildlife dictate that you stay far away from them, don't feed them, don't provoke them, and pack away your trash, among many other rules (be sure to always check at the visitor center to be clear about all of the park's policies). But, of course, many humans don't listen to park rules and end up posing a threat to wildlife. While you're out and about in the park, look for evidence of *humans*—trash on the ground (even small things, like the stick from a lollipop), apple cores along a trail, and even toiletries are attractants for wildlife—and dispose of it in a bear-proof manner.

The park relocates three to five food-conditioned bears per year. These situations are typically caused by visitors not properly storing their food or disposing of their trash.

When an animal, like a black bear in SHEN, stops responding to stimuli that would normally be considered threats (like humans or cars), it's called *wildlife habituation*. On the surface, it might seem great to have a bear wander into your campsite, take a snack from your grill, pose for a few pics, and move on. However, it's actually very problematic, both for humans and animals. When black bears start to associate humans with food, they're more likely to approach humans, cars, tents, and the like in search of a meal that's much easier to access than their food in the wild. Even though bears aren't trying to eat humans, they can be dangerous, especially when they feel threatened. Once they start going to humans for food and tasty trash, they won't stop, and sometimes specific bears must be euthanized to protect park visitors.

The great news is that this is *all avoidable*. If park visitors do simple things, like throwing garbage into a bear-proof container, the bears will keep their distance. Just remember to keep wildlife afraid of you, and don't try to engage with them like you would with a tame animal. By staying at least 50 yards away, you're helping to ensure that bears stay wild for years to come.

**It's possible that you'll see black bears anywhere in the park, so be prepared.**

# GEOLOGY

## TALUS ABOUT IT

**THE APPALACHIAN MOUNTAINS, HUMBLY ERODING** right before our eyes, used to be much more impressive than they are today. At one point, between 350 million and 300 million years ago, they may have been similar in stature to the modern Himalayas. Don't struggle too hard thinking about that time scale—it's nearly impossible for us to understand. Instead, simply take a look at the rock features in SHEN and consider that they have all felt the effects of erosion.

As you're cruising along Skyline Drive (which you should definitely do; it's one of the best park roads out there), pull off at any of the overlooks and take a gander at the mountains around you. Aside from the noteworthy blue haze synonymous with this portion of the Appalachians, notice that these mountains don't have peaks; instead, they are rounded at the top. They are also nowhere near as high in elevation as the Himalayas or even the Rocky Mountains in the western United States. This is thanks to millions and millions of years of erosion.

While you're looking at the mountains in the distance, look for piles of loose rocks, called *talus*, that seem like they've fallen down along the slopes. There are some wonderful expanses of talus, called *talus slopes*, in the park, and once you know what you're looking for, they'll be hard to miss. Talus occurs in other mountain ranges as well, but in SHEN it's remarkably easy to access from a variety of trails, giving you a close-up view of the area's geologic past. Here, talus is the result of activity that occurred during the last few million years. Just like in GRSM, massive glaciers never reached present-day SHEN. However, the freeze/thaw cycles that happened as a result of the glaciers nearby created talus. Basically, water found its way into cracks in rocks and boulders and as time passed, through repetitive freeze/thaw cycles, those chunks split off and tumbled down the mountainside. Why? Because frozen water expands. Look for piles of somewhat evenly sized rock.

Keep an eye out for plants and animals living in these talus slopes—such unique environments are prime homes for a variety of flora and fauna. There are lichens that live here (watch where you step because they take a long time to take hold and grow!) as well as wolf spiders that look similar to lichen on the rock. In the crevices, you might see (or hear!) rattlesnakes and skinks. If you're lucky, you might see the endangered Shenandoah salamander (*Plethodon shenandoah*). It is 3 to 4 inches long as an adult, predominantly dark brownish, sometimes with a narrow red stripe along its back. It is found only on north-facing talus slopes on three mountains in the park: Hawksbill, The Pinnacles, and Stony Man. As always, if you see any wildlife, don't attempt to handle it or do anything to make it notice you.

Talus slopes: Come for the geology, stay for the plants and wildlife!

**Easily accessible talus slopes include the Blackrock Summit trail and the beginning of Hawksbill Gap Trail. More experienced hikers can complete the full Hawksbill Gap Loop, which includes several more talus slopes. Many other talus slopes are visible from park roads and overlooks.**

STATE
**FLORIDA**

ANNUAL VISITATION
**597,000**

YEAR ESTABLISHED
**1947**

# everglades

## a subtly sublime & utterly unique tropical wetland

| | |
|---|---|
| SUPERLATIVE | EVER is the flattest national park |
| CROWD-PLEASING HIKES | Anhinga Trail (easy); Nine Mile Pond canoe trail (moderate); Wilderness Waterway canoe trail (strenuous) |
| NOTABLE ANIMALS | Anhinga (*Anhinga anhinga*); West Indian manatee (*Trichechus manatus*); American alligator (*Alligator mississippiensis*) |
| COMMON PLANTS | Sawgrass (*Cladium jamaicense*); butterfly orchid (*Encyclia tampensis*); royal palm (*Roystonea elata*) |
| ICONIC EXPERIENCE | Trekking to a cypress dome on a ranger-led Slough Slog! |
| IT'S WORTH NOTING | A number of the park campsites, such as Picnic Key, can be reached only by canoe or kayak. |

**WHAT ARE THE EVERGLADES?** Despite being the third-largest park in the lower 48—behind only DEVA and YELL—EVER protects just a small portion of the vast tropical wetland wilderness known as the Everglades. EVER includes some of the richest plant and animal habitats anywhere in the country; it was the first park to be established for its biological diversity rather than its outright scenic value. Don't let that fool you, the subtleties of the landscape here are just as mesmerizing as (and perhaps even more bewildering than) postcard views of mountains and canyons. EVER is an unforgettable wilderness experience for those willing to get their feet wet.

STAMP

PASSPORT

HERE

# GEOLOGY

## WHERE MOLEHILLS ARE MOUNTAINS

**THE LAND UNDER THE SURFACE** of EVER, like most of south Florida, is essentially one long, very gently sloping sheet of limestone. For most of its 150-million-year history, this land was the ocean floor. Over time, calcium carbonate from the shells of tiny marine creatures was compacted and lithified into a deep, flat limestone base. Only in the past few thousand years has this lime rock made its way up to the surface, and even then just barely.

The flat landscape doesn't offer much in the way of topographic relief; elevations in the park range from sea level to just 8 feet higher. Looking for points of geologic interest in EVER requires adjusting your expectations. There aren't huge boulders or road cuts slicing through colorful rock layers. But whereas you would have to climb a thousand feet or more to see changes in a typical mountain ecosystem, stepping up or down *a matter of inches* is enough to change everything in EVER. Here, molehills really can be mountains.

In places where the limestone rises high enough to remain dry all year long, forested island oases—known locally as *hammocks*—of trees grow. Stroll under the dense tree canopy in a hardwood hammock and you will instantly feel the temperature drop. Look for temperate species, like red maple (*Acer rubrum*), rubbing elbows (branches?) with tropical species, like the gumbo limbo (*Bursera simaruba*). Take a minute to breathe in the humid air—it smells much funkier than in a northern forest.

The red, peeling bark of gumbo limbos looks a bit like sunburned skin, giving rise to its nickname: the tourist tree.

Pine rocklands often surround the hammocks. These sunny, open forests of slash pines (*Pinus elliottii*) are high enough to stay dry most of the year. Watch your step as you explore these forests—the ground is an obstacle course of sharp limestone pinnacles. Look out for holes too. Over time, rain that falls on limestone becomes weakly acidic as it moves through layers of decaying vegetation. The acid dissolves the limestone, leaving behind often water-filled depressions, called *solution holes*. These voids in the rock might start off small and shallow, but they continually enlarge over time.

In lower, wetter habitats in the park, solution holes sometimes host clumps of bald cypress trees (*Taxodium distichum*) called *cypress domes*. Take a moment to observe one from the outside. They're called domes because the tallest trees grow in the deepest water at the center of the hole, with shorter trees around the periphery. Cypress domes are often the only places with reliable water during the dry season, which lasts from roughly December through April, making them a magnet for animals seeking food, habitat, or hydration. Alligators sometimes deepen the holes by thrashing around to remove accumulated plant debris. In places where they, or other natural processes, have made the holes especially deep, the dome becomes a doughnut shape, with wetland plants or open water in the "hole."

In places unable to support cypress trees, flooded marshes of sawgrass prairie take over. A few inches lower still, and closer to the coast, mangrove forests thrive, unbothered by the saltwater that discourages most vegetation.

EVER contains the longest unbroken stretch of protected mangrove forest in the Western Hemisphere.

The geology of EVER is not as dramatic as mountain peaks or canyon walls, but has produced wondrous diversity that requires bug spray and soggy feet to explore. Here, the margins of high and low are close together, but worlds apart.

**The Pine Island, Flamingo, and Shark Valley areas feature many short interpretive trails that wind through various habitats. Some limestone outcrops are visible in the Rocky Glades portion of the park.**

# WILDLIFE

## JUST JAWING AROUND

Alligators famously carve out "gator holes" in EVER's flooded prairies and cypress domes. Many of these holes, which provide crucial refuge for other animals during the dry season, have probably been maintained for hundreds of years by successive generations of alligators.

**ALLIGATORS AND CROCODILES HAVE MUCH** in common. They're both members of the crocodilian family. They both survived the extinction event that wiped out the non-avian dinosaurs. Both lay eggs, and in both species the sex of the resulting offspring is determined by the temperature of those eggs. Both have five toes on their front feet and four on their back. Kind of charming, right? One thing they don't often share is location. Only in south Florida, including EVER, do these animals coexist naturally in the wild.

American alligators (*Alligator mississippiensis*) inhabit quite a bit of the southeast United States, hanging out in freshwater sloughs, lakes, and flooded prairies. Southern Florida is at the southern extreme of their range. The opposite is true for American crocodiles (*Crocodylus acutus*), for whom southern Florida is the extreme northern tip of their range. They are the saltwater to the alligator's freshwater, spending their time in coastal areas and in brackish streams where the two waters mix. Alligators are known to sometimes venture into brackish territory, but because they lack the specialized glands that crocodiles have to process and remove excess salt, alligators must return to freshwater before too long.

You will almost certainly see alligators during your time in EVER—they are a regular presence in many areas of the park. Crocodiles are somewhat more secretive, but by no means absent. The easiest distinction between them is jaw shape: Alligators have a rounded, U-shaped jaw, while crocodiles' snouts are pointed and V-shaped. Both are capable of exerting enormous force on whatever they chomp, but the different shapes partly reflect differences in diet. The broader, shovel-shaped alligator jaw is built to withstand the stress of crunching

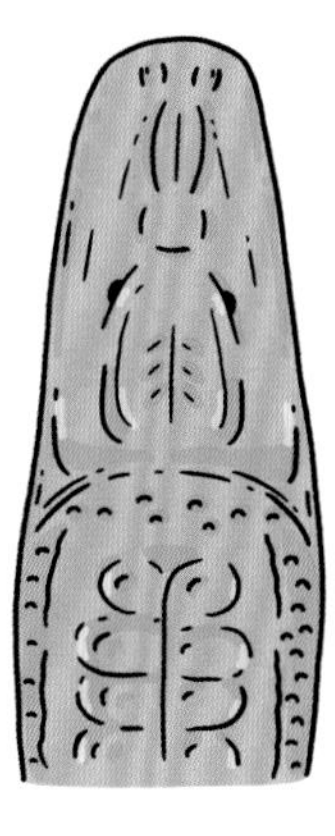

Alligator

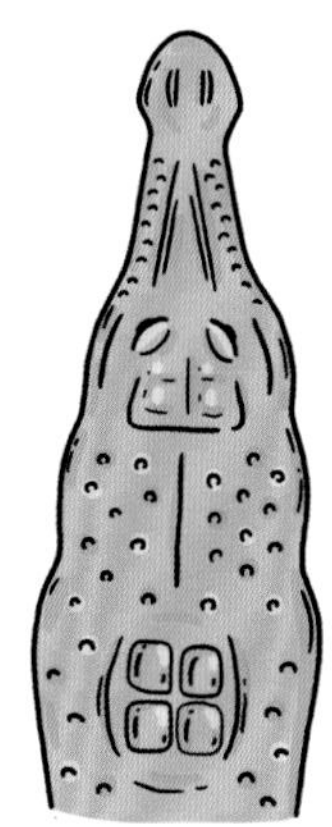

Crocodile

through the hard shells and bones of turtles, birds, fish, and aquatic invertebrates that make up much of their diet. Crocodiles will eat almost anything that moves—fish, turtles, snakes, and small mammals—so their jaw can be thought of as a more generalized weapon. Can you see a bottom tooth sticking up from closed jaws? That's a crocodile—the fourth tooth from the front on the bottom jaw is always visible.

Color can be a somewhat reliable indicator of species. Adult alligators are typically olive, brown, or gray, though they can be nearly black. Their coloring is largely determined by habitat. Alligators who live in algae-filled water tend to be greener; those who live in shadier, tree-covered environs are generally darker. They have cream-colored bellies regardless of habitat. Crocodiles are typically gray-green or brown with a lighter white or yellow underside.

The unique habitat overlap of alligators and crocodiles in EVER is just one more reason that this is such a special place. Remember to treat these reptiles, and all plants and animals, with respect. Both species would rather flee than attack, but they are wild animals that will defend themselves when provoked. Look for the jaw shape and ecosystem differences, and if you get confused, just remember: One of these creatures you'll see later, the other you'll see after a while.

As cold-blooded reptiles, both alligators and crocodiles sun themselves to regulate body temperature. They often look like statues, but they definitely aren't fake! Sunning with the jaws open is an effective way to warm up, not a sign of aggression.

The dry season from December to April is the best time to spot wildlife of all stripes, as the dwindling water forces them into a smaller area of the park than when water is plentiful. Look for alligators near lakes, ponds, canals, and ditches year-round. Crocodiles stick mostly to saltwater and can be found only in the Flamingo area.

# BOTANY

## LIVING ON THE AIR

See page 10 for more about epiphytes in OLYM!

**DURING YOUR TIME IN FLORIDA,** you're more than a little likely to interact with a bromeliad. You'll see them along the boardwalks, near parking lots, and sprinkled through coastal habitats. At the very least, you might sip a drink garnished with the most famous bromeliad: pineapple. That tasty treat doesn't grow in EVER, but plenty of these cool, clingy plants do. You can spot many of Florida's eighteen native bromeliad species throughout the park. But don't look down; all of EVER's bromeliads grow on the trunks and branches of trees. They are epiphytes, plants that grow on other plants without harming them.

Bromeliads come in a huge variety of shapes, sizes, and colors, but all share the basic characteristic of leaves arranged in a spiral pattern around the base. Many look very similar to the severed top of a pineapple. Since their roots never touch the ground, the epiphytic bromeliads of EVER have come up with a few ingenious strategies for getting the water and nutrients they need.

The vast majority of the bromeliads you're likely to encounter in the park are members of the genus *Tillandsia.* Their silvery gray leaves are a great place to get to know another bromeliad trait. Grab your loupe or magnifying glass and have a close (respectful) look at a leaf. You'll quickly see that the surface is covered in tiny cuplike structures. These are trichomes—bromeliads' ingenious way of capturing water from rain, fog, and dew. The trichomes funnel water to the plant and also help reflect sunlight to limit water loss and regulate temperature.

Trichomes aren't the only way bromeliads stock up on water. Some, aptly termed *tank bromeliads*, also grow overlapping leaves at their base to create a reservoir that stays full even during the dry season. Pooled water is sure to attract insects and animals, and tank bromeliads are no exception. Peer inside the base of one and you may very well spot mosquito larvae, centipedes, spiders, or even a tree frog.

## KNOW YOUR BROS

You can spot epiphytic bromeliads on trees, telephone poles, and more throughout the park. Here are five that you can reliably see.

**Reflexed wild pine (*Tillandsia balbisiana*)** Often grows with the cardinal airplant. Look for ants around its bulbous base—sometimes they will defend their territory aggressively from plant-eating animals.

**Powdery strap airplant (*Catopsis berteroniana*)** Quite possibly carnivorous—the powdery substance around the base creates a slippery surface that may help the plant catch insects. This is from a different genus than most bromeliads in EVER; it is found mostly near the coast and has bright green leaves.

**Spanish moss (*Tillandsia usneoides*)** Looks very different from other bromeliads, because it is! The strands are actually thousands of interconnected individual plants living together.

**Cardinal airplant (*Tillandsia fasciculata*)** One of the more common species. Get up close to see small purple flowers between red/greenish yellow bracts (leaf-like structures). Also look for a line, like a seam or the keel on a boat, on the underside of the leaves.

**Giant airplant (*Tillandsia utriculata*)** The largest of the bromeliads in EVER also tends to grow by itself.

Illegal collecting and climate change are serious threats to these amazing plants. Leave them where you find them!

Some tank bromeliads don't hold water; instead, they flare out at the base to create a hollow chamber. Look closely at the reflexed wild pine bromeliad (*Tillandsia balbisiana*) and you just might see acrobat ants scurrying about. The ants live in the chamber, providing the plant with nutrients from their waste. Bromeliads without insect helpers rely on microbes that break down plant matter or other detritus that falls or is blown by wind onto the plants.

Bromeliads thrive in just about every habitat EVER. Look for their intricate forms gracing trees and shrubs in cypress domes, hardwood hammocks, and mangrove forests. Lone trees in the middle of sawgrass prairies are apt to be festooned with the plants. Even trees planted near parking lots, as well as the occasional utility pole, boast their own bromeliad brethren. Wherever you are in the park, take a moment to let your eyes rest on a dense tangle of vegetation—tucked in there are plants that never touch the ground, but easily take root in the imagination.

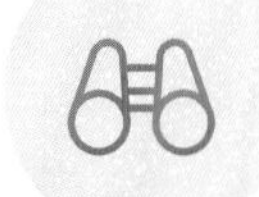

**Bromeliads grow on trees and human-made structures in habitats throughout the park.**

STATE
KENTUCKY

ANNUAL VISITATION
533,000

YEAR ESTABLISHED
1941

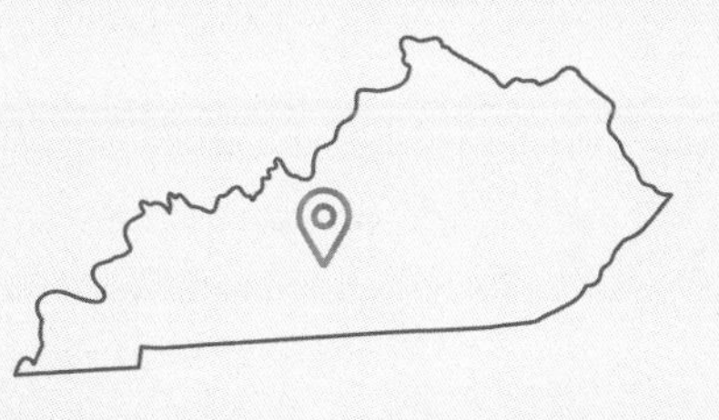

# mammoth cave

## a small park with a big cave & an underappreciated surface

| | |
|---|---|
| SUPERLATIVE | Mammoth Cave is the longest cave system known in the world |
| CROWD-PLEASING HIKES | Sloan's Crossing Pond Trail (easy); Cedar Sink Trail (moderate); Wet Prong Trail (strenuous) |
| NOTABLE ANIMALS | Cave cricket (*Hadenœcus subterraneus*); white-breasted nuthatch (*Sitta carolinensis*); Virginia opossum (*Didelphis virginiana*) |
| COMMON PLANTS | Columbine (*Aquilegia canadensis*); ebony spleenwort (*Asplenium platyneuron*); sycamore (*Platanus occidentalis*) |
| ICONIC EXPERIENCE | Taking a ranger-led cave tour |
| IT'S WORTH NOTING | The park is home to many historic Euro-American cemeteries, which you can visit. |

**WHAT IS MAMMOTH CAVE?** One of the best-known caves in the world, Mammoth Cave is an underground labyrinth that almost defies imagination. To date, more than 400 miles of passages have been mapped, and that may be just the beginning. You can explore the cave on a number of guided tours offered by the park service, but don't neglect the surface world. There are ways to get curious about cave country from up there too. Roads and trails lead to astounding sinkholes and historic sites. MACA is a small park with a whole lot of personality.

# GEOLOGY

## THAT SINKING FEELING

Thousands of years from now, ongoing erosion will turn Mammoth Cave into a series of canyons.

**CAVES, ESPECIALLY BEHEMOTHS SUCH AS** Mammoth, don't form randomly or by chance. They exist in places where underground rock can be dissolved by water in a big way. Geologists call such places *karst landscapes*. Riddled with caves, sinkholes, and springs, karst landscapes make up somewhere between 40 and 60 percent of the land surface in the United States, including large parts of Kentucky.

The limestone hanging out under your feet in MACA makes this quintessential karst country. Left over from an ancient sea, the limestone is now riddled with tiny openings and pores. Slowly but relentlessly, water is dissolving the rock.

A cave is born when rock on the surface resists erosion, but big pockets of rock underground dissolve. That's the story with MACA's eponymous cave. Over time, water flowing underground has eaten away a multitiered, incredibly complex series of passageways through the rock. The hundreds of miles of mapped cave passageways don't stretch out in a tidy line, but instead stack on top of each other like a 3-D underground maze.

But the cave isn't the only wonder here. In some places, limestone dissolves much closer to the surface. The land on top of the newly opened space can stay intact for a while, but eventually it gives way—creating a sinkhole!

Sinkholes can be dramatic, like the one that suddenly swallowed up eight classic Corvettes in Bowling Green, Kentucky, in 2014. (That's a measly 30 minutes south of the park, for anyone keeping score at home.) It's worth noting that if you reside in Florida, you basically live in a giant nascent sinkhole since the entire state is underlain by karst terrain.

Most sinkhole collapses outside of developed areas attract far less attention, but perhaps we should change that. After all, sinkholes are literal windows into cave country, and MACA is riddled with them. Some are a quick walk from the visitor center; others require more exertion. At some, you can peer

## THE BIRTH OF A SINKHOLE

The vast majority of sinkholes in MACA are formed by a process called *cover subsidence.*

Water eats away at rock underground.

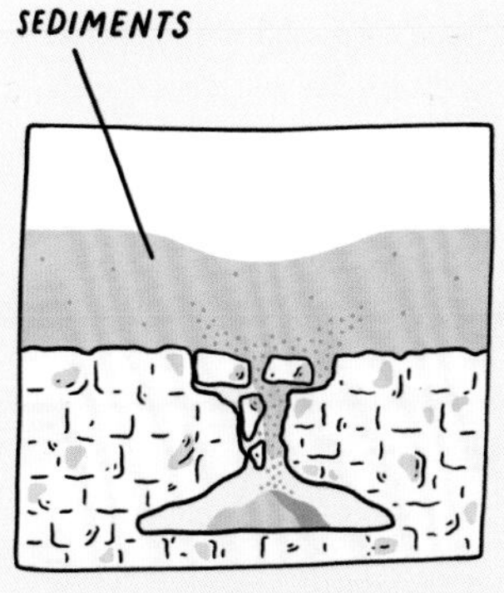


2. Sediments begin to drop down into the empty space underground.

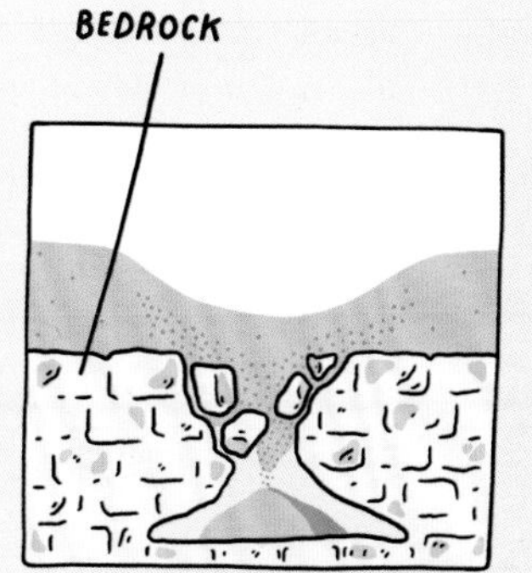


3. More and more sediments and overlying rocks fall into the void; now you can see the sinkhole on the surface.

4. Eventually, the landscape sinks completely into the hole, forming a large divot on the land's surface—a sinkhole!

over the edge; at others, you can take a calf-burning hike down into their depths. From the very large to the smaller, more humble ones, sinkholes are a fascinating way to interact with the world's longest known cave system, without rangers, restless strangers, or cramped spaces.

The sinkholes inside the park are just a tiny fraction of Kentucky's tens of thousands of sinkholes, but they sure are scenic. Some, like Cedar Sink, are huge, with towering walls. Look for debris piles of loose rocks in the bottoms of these pits. You may also see flowing water at the bottom; look for streams that seem to appear from (and disappear just as quickly into) what looks like solid rock. It's not—it's porous, dissolving limestone! Take time to appreciate the rocks exposed on sinkhole walls; these layers are the bones of cave country.

Water in springs and rivers is the active agent in shaping the landscape you are on—and inside of!—while you're in MACA. Places where it emerges from the underground world, or disappears down into it, feel a little magical and a little mysterious. Springs pop up throughout the park as exit holes for some of the water winding its way through underground rock. Some springs, like Turnhole Bend, are colored a deep blue by their mineral composition.

Sinkholes, springs, and caves are classic features of karst landscapes. In this corner of Kentucky, the karst terrain is home to agricultural lands; this small, intriguing national park; and an immense aquifer of pure, tasty spring water. Perhaps a new motto for MACA could be "Come for the cave, stay for the karst terrain." And consider adding a stop to your drive to or from the park, because Kentucky's karst landscape, and its aquifers, are responsible for all that delicious Kentucky bourbon. Cheers to geology!

**Sinkholes and springs litter the park. Some popular ones include Sal Hollow Trail, Turnhole Bend Nature Trail, Mammoth Dome Sink Trail, and Cedar Sink Trail. Many trails near the visitor center show off the region's karst topography.**

## WILDLIFE

### BELEAGUERED BATS

**IF YOU TAKE A TOUR** of Mammoth Cave, you'll get front-row access to creatures living the majority of their lives underground. Cave crickets, millipedes, and salamanders are all standard cave dwellers that you might get to see. You might also see bats, our flying mammal relatives known to hang out (literally) in caves around the world.

Eight species of cave-dwelling bats live in the park, and all used to thrive here until somewhat recently. A fungus called white-nose syndrome (WNS; *Pseudogymnoascus destructans* or *Pd*) has catastrophically reduced bat populations of some species in North America. It was discovered in New York state during the winter of 2006–2007 and is killing millions of bats and spreading rapidly. Unfortunately, its spread can be exacerbated by humans, who aren't harmed by the fungus but can carry the spores from one cave environment to another.

Put simply, WNS is a cold-loving fungus that presents as a white, powdery substance on the nose, wings, and tails of the bats. It disorients bats during hibernation, causing many to leave the safety and warmth of the colony and die. Even those that remain often perish. Though not all bat species appear to be susceptible, many of those that are show little or no ability to fight the infection. Many sites report a fatality rate of 90 to 100 percent for affected populations. In MACA, the population of the once-common little brown bat (*Myotis lucifugus*) has been reduced by 92 percent since the disease was found in the park in 2013.

Bats, although they have a less-than-stellar reputation among many humans, do a lot of great work for us as pest managers. One bat can eat 3,000 to 5,000 insects in a single night, which is hugely helpful in keeping insect populations from exploding and becoming unmanageable. Managing pests naturally—by keeping bats in the environment rather than using pesticides, for example—ensures biodiversity and maintains a healthy ecosystem.

Bats also play a role in pollination!

We have every incentive to try to slow the spread of this disease while we work on a way to stop it altogether. Every national park manages its biosecurity differently, but they all agree that the fungus' spread is inevitable. If you go into Mammoth Cave on a tour, it's mandatory that you participate in their biosecurity efforts. Since the cave is known to have WNS, you'll decontaminate only *after* your tour, which ensures you won't transport the fungus to another cave. The process is simple (and kind of fun!); it simply involves walking across a soapy mat after you leave the cave. There are also restrictions about the gear you're allowed to bring into the cave, so be sure to check out all policies in the park (and any other caves you'd like to visit) before you head out.

**Due to the decimating effects of WNS, seeing bats in the park is fairly rare. Cave tours don't enter areas where bats hibernate, but bats sometimes fly through toured areas. Keep your eyes peeled for them at dusk aboveground as well.**

biscayne

STATE
**FLORIDA**

ANNUAL VISITATION
**469,000**

YEAR ESTABLISHED
**1980**

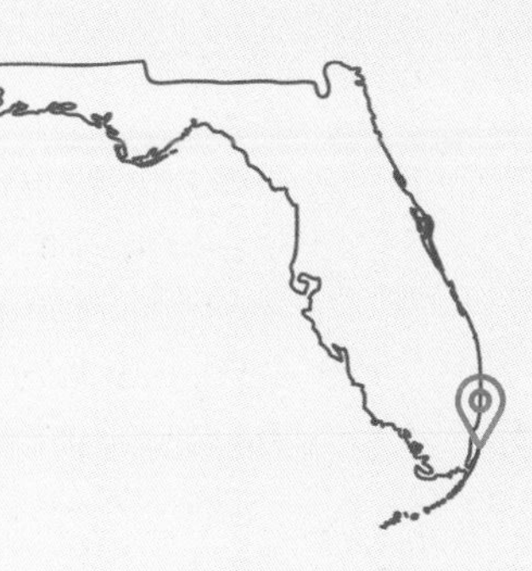

## a watery tropical wilderness in miami's front yard

| | |
|---|---|
| SUPERLATIVE | BISC is the largest marine park in the national park system |
| CROWD-PLEASING HIKES | Spite Highway Trail (easy; bug-covered land); Jones Lagoon (canoe/kayak); Hurricane Creek (canoe/kayak) |
| NOTABLE ANIMALS | Atlantic bottlenose dolphin (*Tursiops truncatus*); brown pelican (*Pelecanus occidentalis*); upside-down jellyfish (*Cassiopea* spp.) |
| COMMON PLANTS | Red mangrove (*Rhizophora mangle*); cabbage palm (*Sabal palmetto*); turtle grass (*Thalassia testudinum*) |
| ICONIC EXPERIENCE | Visiting the Boca Chita lighthouse |
| IT'S WORTH NOTING | You can explore six different shipwrecks along the park's Maritime Heritage Trail. |

**WHAT IS BISCAYNE?** Biscayne Bay, just south of Miami, is a vast, shallow bay covered by waving meadows of sea grass in sparkling, clear water. BISC protects not only the bay but also a rich coastal mangrove forest, the northernmost Florida Keys, and a scattering of coral islands that stretch out into the world's third-largest reef tract in the world. More than 90 percent of the park is beautiful, aquamarine water, so bring a boat, rent something to paddle, or sign up for a guided tour or cruise—BISC is a watery wonder world that can delight and entrance.

# BOTANY

## SEEING SEA BEANS

**MOST OF THE SHORELINE IN** BISC is made of sharply eroding pinnacles of ancient coral reef, but here and there short stretches of sand emerge from the waves. If you pay attention and don't mind getting your hands a little dirty, these small beaches sometimes offer up chances to find a natural message in a bottle from a distant place—a sea bean.

Sea beans are the seeds and seedpods of a diverse number of tropical plants—often woody vines and trees—many of which have a tough, polishable exterior. The hard casing protects a viable seed and some sort of air bubble or other buoyancy device that allows it to float. The beans that wash ashore in BISC are mostly long-distance travelers from Central and South America. Rivers and periodic floods wash them from rain forests into the ocean, where they are picked up by the Gulf Stream, and a lucky few are swept toward Florida's shores. Look for them tucked in among soggy strands of sargassum and sea grass washed in above the tideline.

Some of the most common sea beans in BISC are sea coconuts (*Manicaria saccifera*), which come from a palm with 30-foot leaves, native to Central and South America. You can sometimes find them still in their pods of three that resemble the club symbol on playing cards, though you are more likely to find individual seeds worn smooth from tumbling in the waves. They look a little like a rounded chestnut.

Another common bean is what English speakers call "blisterpod." Those with some Spanish and a sense of humor will appreciate the bean's other common name: *cojones de burro*. (The plant's genus, *Sacoglottis*, doesn't help stifle the giggle.) Look for oblong, slightly wrinkled seeds that average an inch or two long.

A *Manicaria saccifera* palm, the source of the sea coconuts that wash up in BISC.

Depending on how long they've been floating, parts of the bean may have decomposed, revealing the many small seed cavities inside it.

Tropical (or West Indian) almonds (*Terminalia catappa*) are another common sea bean in BISC. These tree seeds resemble oversize, unshelled almonds. Though the trees are originally native to Southeast Asia, Northern Australia, and Madagascar, they have been known to take root and grow in Florida's tropical climate.

You stand a good chance of being able to find two other beans popular with sea bean aficionados: sea hearts and hamburger sea beans. Sea hearts (*Entada gigas*), which come from the Amazon and Orinoco River Basins, are named for their heart-like shape. More whimsical are the hamburger beans (*Mucuna sloanei* and *Mucuna urens*), which come from many tropical places. Alternating light and dark bands really do make these beans look like tiny toy burgers.

Though they rarely wash up this way, hamburger beans grow in pods that resemble giant pea pods, covered in stinging hairs.

Sea beans, of course, are not the only things that wash ashore. They are often tangled in sea grass and other marine plants. Sometimes shells are mixed in as well. Unignorable in all this is the trash. Garbage from park visitors and from all over the world finds its way into ocean currents and onto beaches, including here in BISC. So while you're out admiring the sea beans, pick up a few pieces of trash too—the next sea beaner will certainly thank you for it.

**Elliott Key has the largest beaches in BISC, and Boca Chita Key also includes a small beach. Guided tours to both islands are offered periodically. Other stretches of sand are scattered throughout the park's waters—paddlers and boat owners are free to find and explore them. Do your homework before pocketing any bean; visitors are allowed to remove any exotic species, but they must leave native beans alone.**

## JELLYFISH: SNACKING ON SUNLIGHT

**AS YOU PADDLE** through the shallow water of the bay and calm side channels, you might notice strange creatures carpeting the bottom. They look like plants—and parts of them share some similarities—but what you're actually seeing are jellyfish. The upside-down jellyfish (*Cassiopea frondosa* and *Cassiopea xamachana*) rests its bell on the seafloor and points its tentacles toward the sun to feed photosynthetic algae that live in its tissues. Sunlight feeds the algae, algae feed the jellyfish, everyone's happy. Their sting (actually a cloud of stinging mucus) is relatively mild, but give them—and all plant and animal life—the proper distance and respect they deserve.

*Cassiopea frondosa*, the upside-down jellyfish

STATE
SOUTH CAROLINA

ANNUAL VISITATION
146,000

YEAR ESTABLISHED
2003

congaree

## towering trees dance above meandering waterways

| | |
|---|---|
| SUPERLATIVE | CONG contains the highest concentrations of champion trees (the largest trees within a particular species) in North America |
| CROWD-PLEASING HIKES | Boardwalk Trail (easy); Oakridge Trail (moderate); Cedar Creek Canoe Trail (canoe/kayak) |
| NOTABLE ANIMALS | River otter (*Lutra canadensis*); hairy woodpecker (*Dryobates villosus*); synchronous firefly (*Photinus carolinus*; see page 241) |
| COMMON PLANTS | Water tupelo (*Nyssa aquatica*); American elm (*Ulmus americana*); butterweed (*Packera glabella*) |
| ICONIC EXPERIENCE | Paddling down Cedar Creek |
| IT'S WORTH NOTING | Kayaking through the flooded forest is a unique and unforgettable experience. |

**WHAT IS CONGAREE?** The Congaree were a Native American group decimated by European colonizers; their descendants are part of the modern Catawba Indian Nation. CONG is a forested floodplain, inundated with water about ten times every year. The park is home to some of the tallest trees left in the eastern United States; these grow to an immense size thanks to the rich nutrients brought by flooding and the warm climate. Also abundant here: mosquitos. A warning meter in the visitor center goes from 1 (mild) to 6 (war zone). Bring bug protection and something to paddle, and CONG will treat you to a unique adventure among the trees.

# WILDLIFE

## SLIDERS, COOTERS, AND SNAPPERS

Along with tortoises, turtles are the oldest living group of reptiles. Both groups have survived, with minimal changes, for more than 200 million years.

**YOU DON'T HAVE TO SEARCH** very far in CONG to spot a few animals that you might at first overlook: turtles. They lead mostly quiet lives along the park's lakes, sloughs, creeks, and puddles, munching on plants or insects and contentedly basking in the sun. But don't let their slow life fool you, turtles are ancient masters of evolutionary origami.

Two of the most common turtles in CONG look pretty similar. Yellow-bellied sliders (*Trachemys scripta*) and eastern river cooters (*Pseudemys concinna*) are often found basking on rocks and logs beside rivers and streams. Both have olive-green to brown shells and heads adorned with yellow stripes. Look for wide, well-defined stripes directly behind the slider's eyes. You can also look for two or more solid black blotches just under where the top and bottom halves of the shell meet. Cooters tend to be a bit larger—up to a foot long—and their heads seem a little small for their shell size. Look for yellowish C-shaped markings on the scutes (individual shell plates) toward the front of the shell.

Sharing the water, but rarely seen out of it, is the common snapping turtle (*Chelydra serpentina*). These big hunks are the largest freshwater turtles here and usually weigh between 10 and 30 pounds (about the size of a small dog). They don't bask on logs or rocks like their compatriots. Instead, look for them floating near the water surface when they need to warm up. They spend most of their time in the water, so keep an eye out for older turtles whose shells have become festooned with algae and mud. Adept swimmers, snappers are generally docile in the water. On land they respond more aggressively, since they lack the speed to flee or the ability to fully retreat into their shells. Their namesake bites—which can easily break human skin, but not bone—are reserved for prey (like frogs, birds, and snakes) and humans lacking common sense.

Away from permanent bodies of water, keep your eyes peeled for eastern box turtles (*Terrapene carolina*). These smallish (around 6 inches long) terrestrial turtles can be spotted near many trails in the park. Their high, domed, dark-colored shells are covered in yellow, orange, or olive markings that fade as they age. They can be found far from water sometimes, but they love a good soak in a puddle after a rainstorm.

The one thing all these turtles have in common is a body plan almost too bizarre to be believed. While other animals have evolved hard coverings to protect themselves—beetles have exoskeletons, armadillos have bony plates in their skin, oysters build hard shells from minerals in salt water—turtle shells are something else entirely. A turtle's shell is actually made of its ribs—the bones have broadened and flattened and partially fused to the spine. Turtle shoulder blades, unlike those of virtually every other shoulder-bearing animal (including you), sit inside the turtle's ribcage. The transition from an animal with ribs to one with ribs-turned-to-shell is perhaps one of the most intricate pieces of evolutionary origami yet uncovered.

Turtles are fused to their shells, so contrary to the classic cartoon trope, a turtle can't remove its shell.

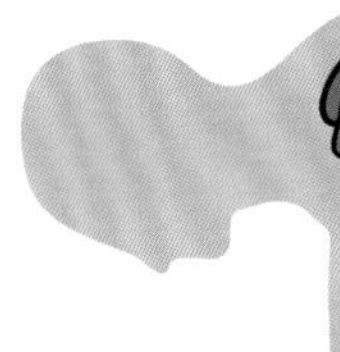

Human

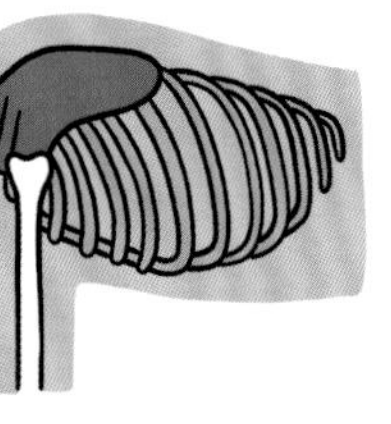

Turtle

Spend time near a creek or a trail in the park and watch the local turtles go about their slow, quiet business. As you watch, lay a hand or two on your ribcage (or your shoulder blades) and think about the incredible evolutionary leaps required to fold and morph bony pairs of organ protectors into a unique dome of defense.

**Weston Lake, off Boardwalk Trail, is a good place to see sliders and snapping turtles. Cedar Creek boasts large numbers of sliders and cooters, often sunning themselves in open areas. Box turtles are common on many trails, but especially so along Bluff Trail that skirts the edge of the floodplain.**

## and another thing

## BALD CYPRESS: TALL, COLORFUL, AND MYSTERIOUS

**MOST CONIFERS ARE** evergreens, but the needles of the bald cypress tree (*Taxodium distichum*) join the leaves of their better-known deciduous counterparts in turning shades of orange, red, and brown before finally falling to the forest floor in autumn. Bald or not, these trees are an impressive sight. They can live for a thousand years and grow more than 120 feet tall—especially in the rich soil and protected habitats of CONG. The purpose of the "knees" sticking out around the trees' massive bases is still cause for debate—perhaps they are meant to deliver oxygen to water-soaked roots or to help stabilize the trees in their frequently flooded habitat. No one knows for sure!

# hawai'i volcanoes

STATE
**HAWAI'I**

ANNUAL VISITATION
**1.1 MILLION**

YEAR ESTABLISHED
**1916**

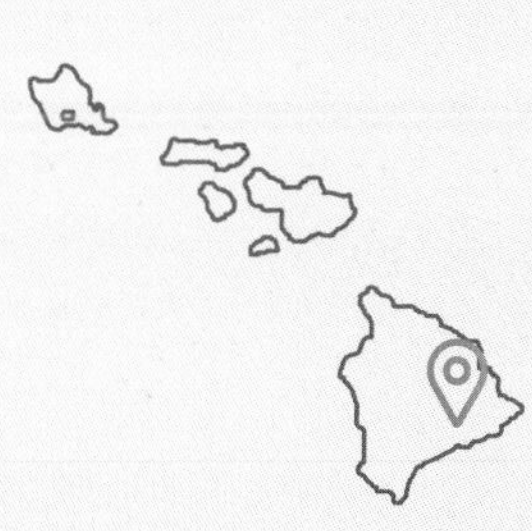

## lush rain forests & fresh lava flows

| | |
|---|---|
| SUPERLATIVE | Mauna Loa is the tallest mountain in the world measured from the sea floor, at 56,000 feet |
| CROWD-PLEASING HIKES | Kīpukapuaulu (easy); Pu'u Loa Petroglyphs Trail (moderate); Ka'u Desert Trail to Hilina Pali (strenuous) |
| NOTABLE ANIMALS | Nēnē/Hawaiian goose (*Branta sandvicensis*); Hawaiian darner dragonfly (*Anax strenuus*); honu/green sea turtle (*Chelonia mydas*) |
| COMMON PLANTS | 'Ōhi'a lehua (*Metrosideros polymorpha*); koa (*Acacia koa*); coconut palm (*Cocos nucifera*; see page 294) |
| ICONIC EXPERIENCE | Finding out that what you came to see has been buried by an eruption |
| IT'S WORTH NOTING | Eruptions that lasted from May to September 2018 have altered some park activities and access points. |

**WHAT ARE HAWAI'I VOLCANOES?** Hawai'i and its volcanoes are inseparable—the entire island chain is a series of volcanoes rising from the ocean floor and above the surface. The island of Hawai'i is home to two incredible volcanoes: Mauna Loa and Kīlauea. Heat, lava, and gasses from those two forces are constantly reworking the surface world here, destroying old habitats and creating new ones. Thousands of miles from any other land mass, Hawai'i's plants and animals are unique—and uniquely threatened. The elements are alive here, and Hawaiian culture is the heartbeat that shows them to the world.

# GEOLOGY

## RAINBOWS IN THE ROCKS

**THE VOLCANOES IN HAVO ARE** among the most active in the world, which means that at any given time it's likely there will be quite a few recently cooled rocks lounging around the place. You would be forgiven for looking out at a field of bare lava and declaring it all to be a flat, perhaps uninteresting black color. But challenge yourself to look closely—there are rainbows in the rocks.

At first glance, the most alluring thing about cooled lava flows might be the endless shapes they take on: smooth and ropy lava, like batter frozen in place, is called *pāhoehoe*; blocky and uneven flows are known as *'a'ā*. All the lava that Kīlauea and the other Hawaiian volcanoes spew has essentially the same mineral composition. The differences in shape and texture you see depend on how sticky or viscous the lava is during the eruption, how quickly or slowly it cools, and even the terrain it covers.

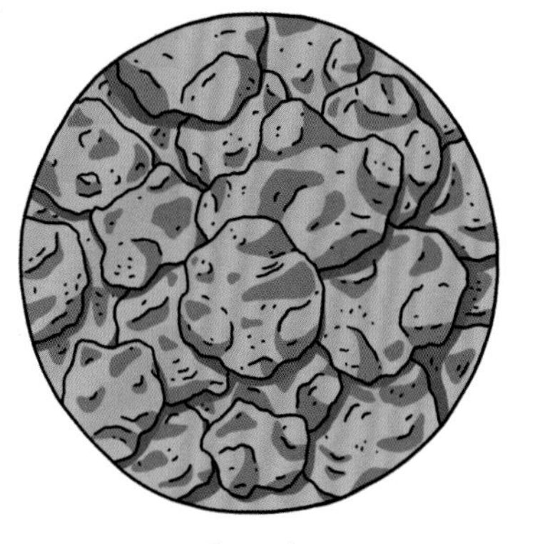

'A'ā (ah-ah)

Pāhoehoe (PAH-hoy-hoy)

The rough texture of 'a'ā flows makes them appear dull, but pay close attention to the smoother surfaces of pāhoehoe flows and you'll soon recognize a world of color. All lava carries bubbles of gas; the amount of bubbles and whether they stretch or burst as the lava flows and cools creates sheens of color in the rocks. Soon after it cools, the outermost layer of a pāhoehoe flow is mostly volcanic glass—lava that cools too quickly for any crystals to form. It's similar to the glass in your house and car, but with much higher levels of iron and magnesium. Thin strands of the glass, known as Pele's hair, are a rich golden brown color. Thicker layers of glass, like the top of a pāhoehoe flow, are virtually black and downright shiny—look for flashes of bluish silver among the black.

In some extremely magical-looking spots, the glassy surface appears iridescent. Shimmering shades of blue, gold, green, and purple make the surface of the flow look like gasoline floating on water. Pāhoehoe flows are not the only places to find iridescent volcanics in the park; look for glimmering colors in glassy spatter and pumice debris near vents as well.

Older lava weathers to shades of brown and orange as the iron in the surface crust reacts with oxygen in wind and rain—yes, the lava rusts!

Lest you think the excitement of volcanic rocks is only surface deep, pull out your loupe or magnifying glass and get respectfully close to that lava! Even without magnification, it's generally possible to make out small mineral grains in the rocks. The most recognizable mineral is olivine, named for its green color. You might also spot flecks of very dark green or black pyroxene, a Greek word meaning "fire stranger." Take a look and a moment to appreciate the strange poetry of some geologic names.

Mineral crystals in lava, big enough for you to see without magnification, begin to form underground as magma slowly cools; the longer magma stays below ground, the larger the crystals can grow.

Volcanic landscapes are dynamic and exciting. Spend some time looking a little more closely; not only will your fellow visitors leave you alone after they spot you out there with your magnifying glass but you will be treated to an almost secret world of color.

**The entire park is volcanic, so just get out there and start exploring.**

## BOTANY

### BLOOMING LAVA

The name of the summit crater on Kīlauea, Halema'uma'u, means "home of the 'āma'u ferns."

'Ōhelo are in the same family as blueberries, cranberries, and huckleberries (see page 131). They look very similar to a poisonous berry, known as 'ākia, that also grows in the park. Do your homework before snacking!

**YOU MIGHT ASSUME THAT THE** lush forests and wide open fields of cooled lava scattered around HAVO don't have much in common. But appearances can be deceiving! Both forests and bare rock are part of an endless cycle of destruction and creation that has been going on here since the first Hawaiian island poked above the waves some 70 million years ago. Plants were the first pioneers to colonize freshly cooled lava then, just as they are today. Hardy and intrepid, early plant colonizers pave the way for successive generations of plants to build up from a thin coating of green to a full-blown tropical forest. Pay attention and you'll quickly learn to spot these forests in the making.

There is no single pattern for how and when plants begin to grow on cooled lava, though there are some broad trends. From tropical sea level to colder mountaintops, rain shadow to rain-drenched, smooth pāhoehoe to blocky 'a'ā, Mauna Loa and Kīlauea are full of microclimates and habitats that influence which plants show up and when. Algae, sometimes called cyanobacteria, are often the first to arrive—usually within a few months of an eruption. They are followed closely by their other cryptogram friends—plantlike organisms that reproduce by spores—like mosses, lichens, and ferns. Whip out your trusty loupe or magnifying glass and have a close look at a few of these ancient creatures growing along the surface of the rocks.

While you're down there, start to notice the abundance of nooks, crannies, and crevices on the surface of the flows. These indentations greatly influence the process of colonization by collecting airborne dirt, water, and even plant seeds and spores. Notice that the rough surface of young 'a'ā flows often hosts a profusion of lichen, whereas the long cracks of ropy pāhoehoe flows often become home to tree seedlings, like those of the 'ōhi'a.

Another important factor for determining which plants establish themselves on flows is the neighbors. While you're

## LAVA LOVING PLANTS

Use this guide to find some of the most common lava colonizing plants in HAVO.

**ʻĀmaʻu fern (*Sadleria cyatheoides*)** Look for distinctly bright-red fronds in young plants; these turn bright green and die back to a stately silver as they age.

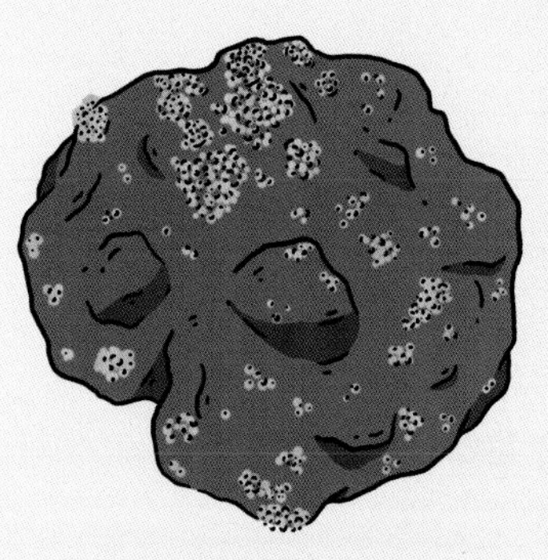

**Lava lichen (*Stereocaulon vulcani*)** Look for stringy gray spots growing on black basalt. The color makes this lichen look like a coating of dust on the black rocks.

**ʻAe fern (*Polypodium pellucidum*)** Look for straight fronds of thick, leathery leaves with smooth edges growing upward from crevices in fresh flows.

**ʻŌhiʻa tree (*Metrosideros polymorpha*)** Look for dark green leaves and brilliant red flowers. ʻŌhiʻa can grow quite tall, but usually appear more shrub-like on new flows.

**ʻŌhelo shrub (*Vaccinium spp.*)** Look for round berries that range in color from red to yellow. These are a staple of the endangered endemic nēnē (Hawaiian goose) and a tasty trailside snack.

standing near a lava flow, look around. The grasses, trees, or other vegetation growing in nearby places that the lava didn't cover have a natural toehold in the area and can move in swiftly when the time is right. Proximity, or lack thereof, has been the defining storyline of life on these islands since they first emerged from the ocean. Two thousand miles of open sea separates these islands from any other landmass, an uncrossable void for most species. Every native life form on these islands was carried here by the wind or the waves, or on another creature that flew, swam, or was blown here. On average, new vascular plants (trees, flowers, vines, and grasses) were able to take hold here only once every 100,000 years.

While native plants still hold the advantage when it comes to colonizing new lava flows, invasive species disrupt the process at every stage. Scientists are still unraveling the impacts of these invaders.

The destruction of a recent eruption can be difficult to overlook, especially for mainlanders who may be less accustomed to the subtleties of coexisting with active volcanoes. As you look out over the lava, both young and old, on display at HAVO, see if you can recalibrate your thinking. Over hundreds or thousands of years, the first plants that gain a hold on newly cooled lava pave the way for larger, less intrepid species. Humble algae and beautiful shrubs are an early step on the way to a life-filled tropical forest. Like forest fires, new lava flows aren't the end of anything; they are the canvas on which another endless cycle of rebirth is beginning.

**The long history of active volcanoes in HAVO makes it an excellent place to explore lava flows both young and old. Chain of Craters Road is a great place to start. Check in with park staff about recent eruptive conditions and suitable hiking trails.**

**HAWAI'I IS HOME TO COUNTLESS** unique animals, thanks to its young geologic age and its isolation. There are only two native mammals—the hoary bat and the Hawaiian monk seal—on the island chain, and its birds and insects can seem like they're from another world. There are hundreds of moth species that call the islands home, and some of those moths' caterpillars are very, very hungry.

Worldwide, the vast majority of Lepidoptera (the order that includes moths and butterflies) caterpillars are herbivores. They simply eat the plants upon which they were born, like the sweet creatures they are. However, in Hawai'i, about fifteen species of pug moth caterpillars (*Eupithecia* spp.) are carnivores. Maybe more interestingly, they are ambush predators. You read that right. Ambush. Predators. These inchworm caterpillars (we know, adorable) are true carnivores, meaning that they don't eat anything else besides other animals. In their case, they usually eat flies, other insects, and spiders. Typically these cute killers are brown or green and look a lot like twigs or stems. Moth life cycles are really similar to butterflies' (see page 36 for more on that), and the adults lay their eggs on plants. The newly hatched caterpillar creates a little spot for itself by chewing away some of the plant, and then it perches. And waits. When an insect lands on the caterpillar, tiny hairs on its abdomen activate the caterpillar's ambush—it springs at the insect, grabs it with its legs, returns to its original position, and enjoys its feast. That's that!

So, *why* are these caterpillars carnivores? It's a pretty significant departure from their caterpillar counterparts, and presumably their ancestors were not carnivores when they arrived. Like some great questions in science, it remains unanswered. Perhaps, as they were colonizing the island,

## WILDLIFE

### THOSE CATERPILLARS EAT WHAT?

For more on Hawaiian evolution, see page 290; about the silversword alliance in HALE.

Pug moth caterpillar perched for attack.

there weren't as many creatures filling the insect-eating niche. Without ants, praying mantises, and other insect predators, these caterpillars had a wide-open field to fill. Another factor may have been a drive to get more protein. Elsewhere, *Eupithecia* caterpillars are herbivores that eat protein-rich portions of plants. The caterpillars that ended up on Hawai'i may have been predisposed to prefer a high-protein diet.

You'll have to go into the forest to try to find these caterpillars, and it's not likely that you'll see them feeding. If you do, most definitely snap some pics and tell park staff because they'll probably be super-delighted about it. You might see the pug moths as adults, which are unfortunately just kind of average-looking brownish moths. Adult Lepidoptera don't typically eat solid food (they simply sip up liquids such as nectar instead), so you won't find those pug moths pouncing on flies. Just remember to ask every moth you see, "Did *you* spend your former life killing bugs?"

**Any hike in a forest in the park offers the chance of seeing a caterpillar—be sure to keep your eyes peeled!**

STATE
**HAWAI'I**

ANNUAL VISITATION
**1 MILLION**

YEAR ESTABLISHED
**1961**

# haleakalā

## brilliant colors & stark expanses

| | |
|---|---|
| **SUPERLATIVE** | HALE is home to more endangered species than any other national park unit |
| **CROWD-PLEASING HIKES** | Hosmer Grove (easy); Pipiwai Trail (moderate); Keonehe'ehe'e (Sliding Sands; strenuous) |
| **NOTABLE ANIMALS** | 'Apapane/honeycreeper (*Himatione sanguinea*); carnivorous caterpillar (*Eupithecia* spp.; see page 285); pueo/Hawaiian short-eared owl (*Asio flammeus sandwichensis*) |
| **COMMON PLANTS** | Āhinahina; (*Argyroxiphium sandwicense* subsp. *macrocephalum*); koa (*Acacia koa*); pūkiawe shrub (*Leptecophylla tameiameiae*) |
| **ICONIC EXPERIENCE** | Watching a sunrise from the rim |
| **IT'S WORTH NOTING** | The inside of the "crater" is one of the quietest places on Earth. |

**WHAT IS HALEAKALĀ?** Haleakalā is the heavily eroded summit of the massive shield volcano that makes up about three-fourths of the island of Maui. Popularly called a crater, it's actually the meeting point of two valleys carved into soft volcanic rock by streams and landslides. HALE sweeps from the colorful and barren world of the summit to the black rock and blue ocean below in the park's coastal Kīpahulu District. Steeped in Hawaiian cultural traditions, HALE is a place that can open your eyes to much of what makes these far-flung volcanic islands truly unique.

# GEOLOGY

## STILL RUNNING HOT

YELL also sits on top of a hot spot.

**IN A FEW PLACES SCATTERED** around the globe, plumes of hot rock rise from deep inside the planet and erupt through Earth's crust. The Hawaiian Islands stretch across one such hot spot. Over the millennia that this particular site has been active, the Pacific Plate has been slowly sliding across it. Like a conveyor belt, the plate has carried along an enormous chain of underwater volcanoes created by the hot spot. The eight islands that make up the State of Hawai'i are the youngest of some eighty volcanoes that stretch for more than 3,600 miles into the North Pacific.

Only the youngest volcanoes—those on the Island of Hawai'i as well as the still submarine Lō'ihi—are still actively growing larger, fed by a direct and intense connection to the hot spot. But that doesn't mean the volcanic theatrics are done on the older islands. Standing on the rim of Haleakalā today, you are looking at a brilliantly colorful reminder that long after the Pacific Plate has whisked them away, the heat from the hot spot lingers on for these volcanoes.

Haleakalā first began erupting underwater around 2 million years ago. It spent a few hundred thousand years erupting layer after layer of lava and building itself above the waves. By about 900,000 years ago, the movement of the Pacific Plate had carried it away from the most intense heat of the hot spot. That's when Haleakalā entered what scientists call the post-shield stage of its volcanism—the chemistry of its lava changed as it came from deeper reservoirs, and the timing of eruptions changed too, with a longer period of time between each eruption.

Today, looking out from the rim—or, if you're feeling intrepid, from inside the crater—what you see are very young rocks. Some, like the cinder cones and lava flows covering the floor, are no more than a few hundred years old. The crater you're looking out at is evidence of the post-shield phase as well. It's not a crater or a caldera formed during an explosive eruption; it is the product of erosion—two huge stream valleys cutting their way into the sides of the volcano until they were so large they joined together.

Evidence from past eruptions tells us that more are still to come on Haleakalā; over the past 10,000 years it has erupted, on average, every 200 to 500 years in this phase of its life. While it's impossible to exactly predict the next eruption, it is possible to confidently declare that Haleakalā isn't done yet. It remains an active volcano that could erupt again.

It's easy to lose yourself in the otherworldly sights and fountains of color visible on Haleakalā; it's a stunning landscape! But take a moment to think about the immense heat that is building the Island of Hawai'i next door. That force of nature is 200 miles away under the Island of Hawai'i, but it's powerful enough to continue sculpting the ground beneath your feet.

The Pacific Plate is sliding over the hot spot at about the speed at which your fingernails grow.

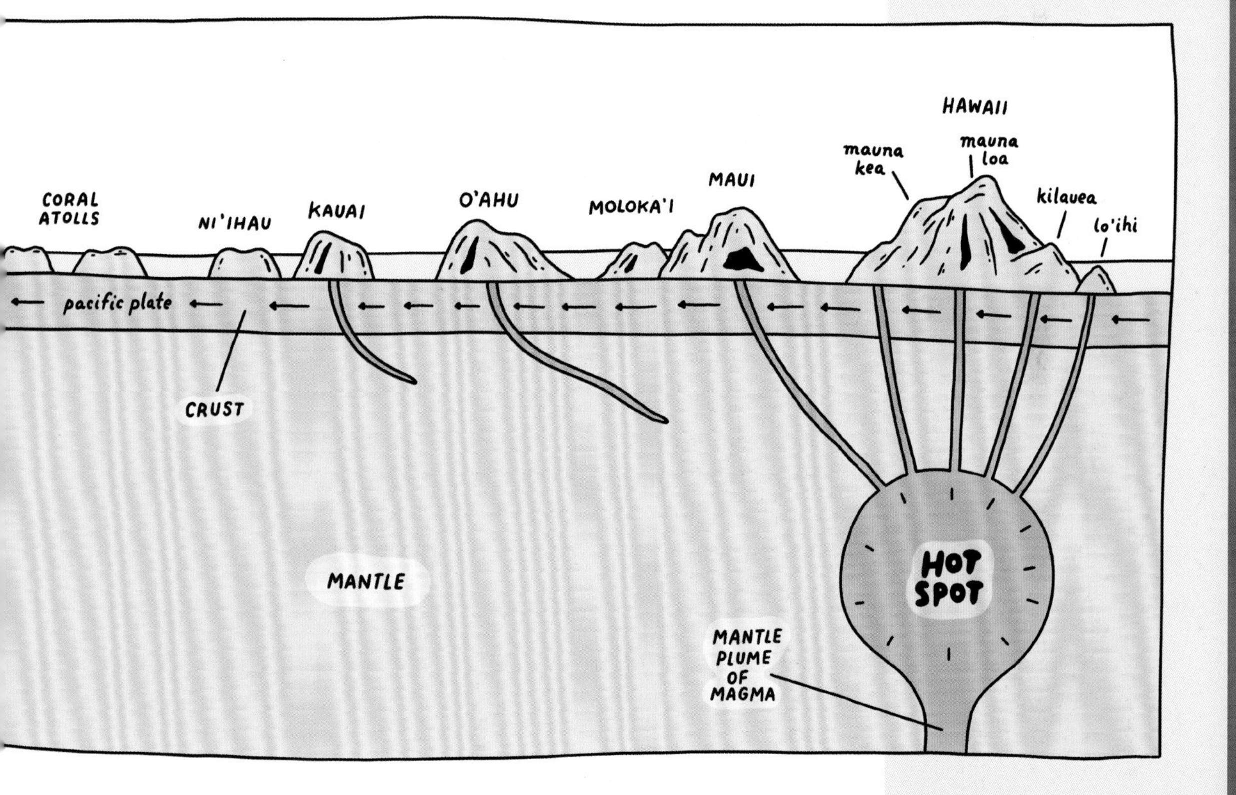


**If you're standing on any of the Hawaiian Islands, you are standing on—and looking at—rocks that came from the hot spot!**

# BOTANY

## AN EVOLUTIONARY ALLIANCE

**HAWAI'I'S ISOLATION AND THE NATURE** of its geologic formation mean that it's full of species of plants and animals endemic to the island chain. The islands popped up relatively recently and are the product of underground volcanism. Ancestors of species native to the islands presumably got there by flying, riding debris on the ocean waves, or tagging along with something that was flying there, or on the wind. It's possible that some even arrived by way of a fortuitous combination of all of these, like riding in on an intense storm.

When the ancestors arrived, the land was fresh and barren. With no other competition, plants and animals could occupy all the habitats available. Some species took hold as they were. Since Hawai'i's islands were so new, there wasn't competition for space, and predators were nonexistent. Offspring of these species went off to the uncolonized areas, and they eventually died off or continued to adapt to new environments.

When adaptations take hold, entirely new species can eventually evolve. This is called *adaptive radiation*, and it's one reason Hawai'i's flora and fauna are so diverse. One great example is the silversword alliance. Sure, it sounds like something from a dystopian young adult novel, but it's actually a group of thirty or so species of plants that evidence suggests are all descended from a single ancestor. It's believed that this ancestor was a tarweed and came over from North America. Members of the silversword alliance are trees, vines, shrubs, and so many plants in between. Since they had the entire landscape to work with, they were able to adapt to every terrestrial environment available. Where water was plentiful, larger trees and vines flourished, and where conditions were harsher or drier, shrubs and rosettes thrived.

The park's famous Haleakala silversword (*Argyroxiphium sandwicense* subsp. *macrocephalum*) is specially adapted to thrive in harsh and dry environments, which is how it lives on the Haleakalā crater. The lower part of the plant is a rosette shape, with fleshy leaves bearing silver hairs.

## MEMBERS OF THE ALLIANCE

There are more than thirty species within the silversword alliance; here are three that you might see during your visit to HALE.

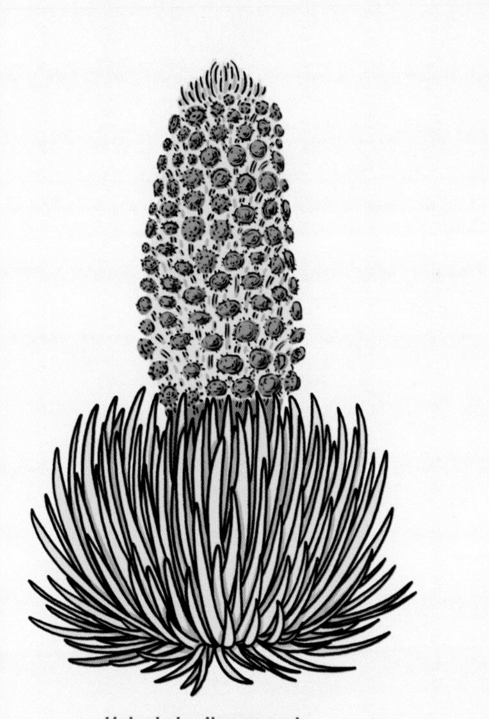

**Haleakala silversword (*Argyroxiphium sandwicense* subsp. *macrocephalum*)** Look for a rosette close to the ground with fleshy leaves and silvery hairs. If you're lucky, you may see one during its full bloom, a stalk up to 6 feet high with rose-purple flowers.

**Mountain dubautia (*Dubautia menziesii*)** Look for a small shrub with vertically oriented branches. Its green leaves are stiff and twisted, and it has small yellow flowers when it's in bloom.

**Greensword (*Argyroxiphium grayanum*)** Look for a low-lying shrub with green leaves growing in a rosette shape. In a bog environment, they grow as dwarf shrubs and don't get taller than 12 inches.

This setup is pretty sweet—the hairs protect the plant from the light and heat they're exposed to at this high altitude (above 10,000 feet), and the leaves store extra water so the plant can withstand the dry climate. It also has a shallow, fibrous root system that helps it absorb moisture from the mist. But the silversword might be best known for its incredible flower display. Once in its life (which can be up to 40 years), the plant launches a stalk from its center, which grows up to 6 feet tall, with a spectacular column of rose-purple flowers. The flowers disperse seeds and then the plant dies.

Unfortunately, these plants are extremely rare because visitors step on their shallow roots and used to (and continue to) uproot them to take home as souvenirs. These native plants are also vulnerable to overcrowding by invasive plants, nonnative ungulates such as pigs and deer snacking on them, and the always-troubling climate change issue. Park staff helps them out by fencing them in, removing invasive species, raising visitor awareness about the fragility of the plant, and collecting seeds to grow in captivity until they can be replanted in the wild. As you're out looking for these amazing members of the silversword alliance, remember to stay on the trail and take photos instead of an entire plant.

**You can see the Haleakala silversword on alpine desert trails, like Keonehe'ehe'e (Sliding Sands). Be sure to stay on trails wherever you hike in the park.**

TERRITORY

# ST. JOHN, US VIRGIN ISLANDS

ANNUAL VISITATION

**112,000**

YEAR ESTABLISHED

**1956**

## jungle-covered hills meet white sand & turquoise water

| | |
|---|---|
| CROWD-PLEASING HIKES | Trunk Bay Underwater Trail (easy); Ram Head Trail (moderate); Reef Bay Trail (strenuous) |
| NOTABLE ANIMALS | Donkey (*Equus asinus*); Caribbean hermit crab (*Coenobita clypeatus*); hawksbill sea turtle (*Eretmochelys imbricata*) |
| COMMON PLANTS | Seagrape (*Coccoloba uvifera*); manchineel (*Hippomane mancinella*); Turk's cap (*Melocactus intortus*) |
| ICONIC EXPERIENCE | Taking a dip in the sparkling clear water |
| IT'S WORTH NOTING | Almost the entire island is covered in second-generation plant growth because nearly everything was clear cut during the horrifying age of slavery and sugar plantations. |

**WHAT ARE THE VIRGIN ISLANDS?** The Virgin Islands are the exposed tips of a submarine mountain range that stretches from Cuba to Trinidad. VIIS covers about two-thirds of the island of St. John, one of the three larger US Virgin Islands. (The others are St. Thomas and St. Croix.) The rugged volcanic terrain of the park includes forest-covered hills, rocky cliffs, and iconic white-sand beaches within protected bays. The park boundaries extend out into the sea where they meet Virgin Islands Coral Reef National Monument—the only national park site without any dry land. From tropical forests to coral reefs bursting with life, VIIS offers nearly unlimited adventures under the bright Caribbean sun.

## BOTANY

### WORLD TRAVELERS & HOMETOWN HEROES

In 2017, Hurricane Irma killed many coconut palms in VIIS. Due to their nonnative status, the park decided not to replant them in favor of native species. They are less dominant than they once were in the park, but still present.

**CLOSE YOUR EYES AND THINK** of a palm tree. Odds are you were picturing a coconut palm (*Cocos nucifera*), a tall, slender trunk topped with bunches of green fronds and tasty coconuts. These trees are ubiquitous in and practically synonymous with tropical places, especially beaches. You'll see them in VIIS whether you're sunning yourself on the sand or hiking through the forest. Despite the ease with which they have taken up residence in the Caribbean, coconut palms aren't native here—or anywhere else in this hemisphere for that matter.

Coconut palms originated in two different places: one genetic group evolved along the coasts of India, the other came from Southeast Asia. Palms in the Caribbean tend to be of the Southeast Asian variety, but very few trees remain purely one thing or the other. Coconuts are the perfect traveler's companion. The fruits themselves are sources of both food and water, while the rest of the tree can be turned into just about everything, from housing materials to oil to alcohol. Everybody from Polynesian voyagers to European colonizers toted coconuts on their journeys, spreading the plants around the world over a few thousand years. It was only recently with the help of genetic studies that scientists were able to rewind the journey.

Despite not being native to the Caribbean, coconut palms occupy an ecological niche that is far from invasive. Sometimes they are referred to as *naturalized*—transported here because of humans but able to reproduce naturally, and so well established that they are part of the ecological and cultural history of this place. It is all but impossible to know when they arrived, but they are definitely here to stay.

Whenever coconut palms did arrive, they began to grow beside at least one species of native Virgin Island palm that you can still see today—the tyre or teyer palm (*Coccothrinax alta*). Tyre palms grow abundantly in the moist forests that drape St. John's northern shore; look for tall, skinny trees that don't flare out at the base like coconut palms do. The easiest way to distinguish a tyre palm from a coconut palm is by leaf shape. Both plants have compound leaves that are made up of many individual leaflets. Tyre palms have leaflets arranged in a fan shape; in coconut palms, the leaflets are arranged in straight lines. The silvery underside of tyre palm leaves sometimes makes them look like they are flashing brightly in the Caribbean sun.

Tyre palms have fan-shaped leaves.

Coconut and tyre palms, like all palms, share some pretty curious growth habits. They grow out rather than up when they first sprout, achieving their maximum diameter before they add height. Throughout their lives, they add new growth only at their tops. Their "trunks" are actually stems that lack many structures present in true trees (broadleaf trees and conifers) and are covered in hardened cells left behind by the bases of fallen fronds rather than bark.

Coconut palms that have sailed the high seas and tyre palms that have grown in the places of their ancestors are two small additions to the riot of green that makes up the forests in VIIS. The palms share some similarities, are distinguished by a few differences, and ultimately help make this place the lush, green tropical world that it is.

Coconut palms have long, feather-shaped leaves.

**Coconut palms can be found throughout the island, especially near the visitor center at Cruz Bay. The wet, forest habitat of the tyre palm can be found along Highway 20 on the northern shore, abutting many of the most famous beaches on the island.**

and another thing

## ISLAND CONNECTIONS: SHALLOW WATER AND DEEP OCEAN

**DURING THE LAST** glacial advance, the scattered islands of Puerto Rico, St. John, St. Thomas, and the British Virgin Islands were all connected by dry land. St. Croix wasn't in on the party because of an 8,000-foot-deep trench on the seafloor that separates it from the rest of its Virgin Island counterparts. That depth is nothing compared to the 30,000-foot drop-off to the north of the rest of the Virgin Islands. Known as the Puerto Rico Trench, bordering the Caribbean Sea, it is the deepest part of the Atlantic Ocean.

# dry tortugas

**STATE**
FLORIDA

**ANNUAL VISITATION**
57,000

**YEAR ESTABLISHED**
1992

## small specks of sand in a vast turquoise sea

| | |
|---|---|
| CROWD-PLEASING HIKES | Tour of Fort Jefferson (easy); snorkeling around the moat wall (moderate); snorkeling or diving the Windjammer Wreck (strenuous) |
| NOTABLE ANIMALS | Green sea turtle (*Chelonia mydas*); masked booby (*Sula dactylatra*); nurse shark (*Ginglymostoma cirratum*) |
| COMMON PLANTS | Manatee grass (*Syringodium filiforme*); sea lavender (*Tournefortia gnaphalodes*); coconut palm (*Cocos nucifera*; see page 294) |
| ICONIC EXPERIENCE | Bird watching! |
| IT'S WORTH NOTING | You can feel like you have the island to yourself and sleep under twinkling stars at the small primitive campground on Garden Key. |

**WHAT ARE THE DRY TORTUGAS?** The Dry Tortugas are a series of small coral islands and sandbars at the very end of the Florida Keys. Named for their abundance of turtles and lack of freshwater, the islands are constantly being reshaped by wild weather and water. DRTO protects the islands and a large part of the shallow, crystal-clear waters that surround them—less than 1 percent of the park is dry land. The second largest island, Garden Key, is dominated by the incongruous sight of a nineteenth-century brick fortress. Though never completed, Fort Jefferson was originally meant to protect US shipping interests in the Gulf of Mexico and also served as a military prison during the Civil War. Wildlife, above and below the water's surface, and the saturated colors of the tropics are the main stories here. Dive in!

# WILDLIFE

## ANIMAL, VEGETABLE, MINERAL

**THE WORDS *FLORIDA* AND *NORTHERNMOST*** don't often find themselves side by side in a sentence. But in the crystalline waters of Florida's south, including at DRTO, coral reefs—iconic tropical structures—reach their northernmost limit. Strap on your snorkel or your scuba tank and get ready to appreciate the tiny creatures that thrive on teamwork and use seawater as building blocks.

About seventy-five species of corals live in the waters off DRTO. The actual family relationships of corals are messy (who among us doesn't identify with that?), but all corals share the same basic body plan. Known as *polyps*, their bodies are made of a central gut cavity surrounded by tentacles and attached to a hard surface. Corals are colonial animals, meaning that polyps grow together in large groups called *colonies*. Individual colonies can be as small as a fist or as large as a car. Groups of colonies form a coral reef.

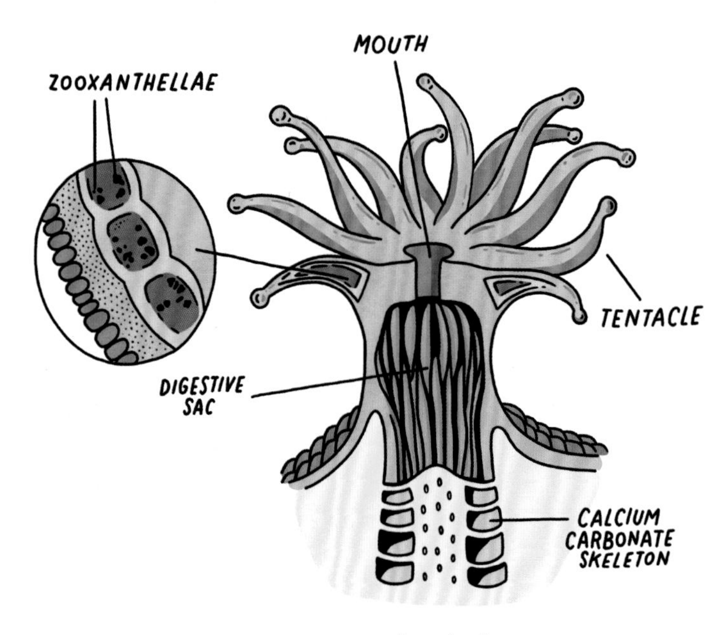


A coral polyp

Coral polyps are soft, squishy creatures, but those that build reefs (the hard, or stony, corals) have a spectacular secret hidden in their gelatinous tissues: they can build solid, stable structures using nothing but seawater. Polyps draw seawater into their tissues and then, by separating out and recombining various minerals and ions from the water, the tiny animals build themselves a rigid "skeleton" out of calcium carbonate.

If you've ever taken medicine for heartburn or indigestion, there's a good chance you ingested a form of calcium carbonate—it's a multitalented molecule.

Polyps don't live alone in their mineral castles—most of the corals you might see under the waves get their vibrant hues from a type of algae living symbiotically with the polyps. Zooxanthellae are single-celled algae that happily share the nutritious rewards of photosynthesis with their hosts in return for the cozy living space inside the polyp. The algae also get some extra nutrients when the polyp's tentacles snare tiny prey as it floats by. Powered by this dual source of nutrients, shallow-water hard corals are able to build their underwater empires.

Want to know more about algae collaborations? Check out "Lovin' Lichens" on page 39.

As you swim over the reef, keep an eye out for sea fans swaying in the currents. They are also corals, but they belong to a different group than their reef-building relatives. These soft corals eschew the work of building a calcium carbonate skeleton. Instead, they are supported by a hard protein similar to a deer antler. They don't build reefs but happily grow within the reef ecosystem.

Corals, like most sea creatures, are facing devastation from human-caused climate change. The amount of carbon dioxide ($CO_2$) that we as a species have pumped into the atmosphere is altering the chemistry of the oceans. Scientists are working to understand the mechanics of it all, but at this point it's clear that changing ocean conditions are at least partly responsible for massive zooxanthellae die-offs (known as coral bleaching) while also making it more difficult for corals to build their homes.

Coral reefs are among the most diverse and visually stunning habitats in the ocean or on land. They provide food and shelter for a truly mind-blowing number of creatures. There aren't many other national parks offering the chance to so closely experience the dynamic and vibrant world of coral reefs found at DRTO. So get in the water and see for yourself! All of the diversity swirling around you is here because a tiny, tentacled creature built a house out of the sea and invited a teeny, tiny alga to move in. Those small acts add up to big sights.

**Coral reefs are abundant in the waters of DRTO and can be accessed by snorkeling or diving. Give the coral as much space as you would any other animal or natural feature; never touch or disturb them in any way.**

## and another thing

## FRIGATEBIRDS: SOARING HIGH AND DRY

LONG KEY, ABOUT a mile east of Garden Key, is home to the only nesting site for magnificent frigatebirds (*Fregata magnificens*) in the United States. Grab some binoculars and have a look at birds that don't often spend time on solid ground. On land, their short legs and webbed feet make walking next to impossible. But in the air, frigatebirds are master fliers who expertly ride air currents, flap their wings only rarely, and stay aloft for weeks at a time. Despite being seabirds, magnificent frigatebirds lack waterproof feathers and so never land on the water.

TERRITORY
**AMERICAN SAMOA**

ANNUAL VISITATION
**29,000**

YEAR ESTABLISHED
**1988**

## rain forests & coral reefs in pacific splendor

| | |
|---|---|
| SUPERLATIVE | NPSA is the only US national park located south of the equator |
| CROWD-PLEASING HIKES | Pola Island Trail (easy); Tuafanua Trail (moderate); Mount ʻAlava Adventure Trail (strenuous) |
| NOTABLE ANIMALS | Raccoon butterflyfish (*Tifitifi laumea*); blue starfish (*Lynkia laevegata*); Micronesian skink (*Emoia adspersa*) |
| COMMON PLANTS | Saʻato/swamp fern (*Acrostichum aureum*); wedgeleaf spleenwort (*Asplenium cuneatum*); ʻatone/Samoan nutmeg (*Myristica fatua*) |
| ICONIC EXPERIENCE | Snorkeling off the coast of Ofu to see corals and marine life (note that you must bring your own snorkeling gear) |
| IT'S WORTH NOTING | The Samoan people have lived here for thousands of years and continue to do so today. Their culture is part of the area, so be sure to learn about it in advance and respect it during your visit. |

**WHAT IS THE NATIONAL PARK OF AMERICAN SAMOA?** Spread across three islands (Tutuila, Ofu, and Taʻū) in the Samoan Archipelago, NPSA is one of the most remote national parks in the system. With its isolation comes relative solitude; whether you hike into the rain forest and walk among trees and ferns or snorkel offshore to see fish and corals, you can count on an experience unlike any other. NPSA's South Pacific location makes it susceptible to cyclones (the South Pacific term for what North Americans know as hurricanes) and tsunamis at any time of year. Plan accordingly to enjoy this island paradise, and take note of how the land can recover after a natural disaster.

STAMP PASSPORT HERE

# WILDLIFE

## FANTASTIC FLYING FOXES

The three bat species are the only mammals native to NPSA.

**BATS IN THE NATIONAL PARKS** are often associated with caves, insect management, and late-night living. They're not known for their great size or incredible soaring ability, but that's not the end of the story. NPSA is home to three species of bats, and two of them are known as "flying foxes" or "fruit bats." These gentle giants have wingspans up to 3 feet; they play an important role in keeping the ecosystem of the islands balanced and are vulnerable to habitat loss and hunting.

Roosting pe'a vao; just imagine a cat wearing a cape, that's basically a flying fox roosting.

There are two unique species of flying foxes in NPSA: pe'a vao, or the Samoan flying fox (*Pteropus samoensis*), and pe'a fanua, or the insular flying fox (*Pteropus tonganus*). Both are active day and night, and it's likely you'll see either species during your time in the park. The two species look really similar to the untrained eye, but there are some ways you might be able to tell the difference. Pe'a vao tends to soar and ride thermals, whereas pe'a fanua tends to flap its wings more frequently in flight. If you see them roosting in trees that also can provide a clue to which species you're seeing: Pe'a vao roost alone, as mated pairs, or as a female with her baby; pe'a fanua roost in large colonies of hundreds or even thousands. Flying fox mothers are known for their caretaking abilities—each one will care for her baby (they usually have only one per year) for several months before it's weaned. They even carry their babies while they're flying!

Differences aside, both species of flying fox play a huge role in supporting the flora on the islands. These bats are vegetarians and rely on the seeds, fruits, nectar, and sap of plants and trees to survive. In fact, even if you don't see the bats themselves, you may see a lot of mushy pulp under trees, like breadfruit trees ('ulu). That pulp might be remnants of the fruit bats' feeding in the trees. Since they do so much flying from plant to plant, tree to tree, or flower to flower, they help those species reproduce. Sometimes they simply carry pollen in their facial hair and on their bodies. They also consume seeds and then disperse them elsewhere in their feces. Perhaps even more impressive, though, is the fact that they're the only animals strong enough to carry tree fruits with larger seeds for longer distances. This delivery service is significant, especially when parts of the island have minimal tree cover due to natural disasters such as tsunamis—these seeds can help start the next phase of regrowth in that area.

Pe'a vao
(*Pteropus samoensis*)

Like bats in other parts of the world, flying foxes have some battles to fight and stereotypes to overcome. They're completely harmless to humans but are still considered pests because they eat fruits intended for human consumption. Flying foxes are often shot with a slingshot or gun; with this predation on top of habitat loss and their low birthrate, efforts will need to continue to ensure that these helpful mammals have a future in NPSA. As you see these gentle giants flapping or soaring in the park, be sure to sing their praises and celebrate the work they do to keep the islands thriving.

**Flying foxes can be found throughout the park—look for them roosting (hanging upside down) from trees, or in the sky, either flapping or soaring.**

and another thing

## NAVIGATING NIGHT SKIES: SOUTHERN HEMISPHERE EDITION

DEPENDING ON WHERE you live, you may be familiar with only the night skies of the Northern Hemisphere and their well-known North Star. Since NPSA is located in the Southern Hemisphere, there is no North Star to guide the way. Of course, humans have been using the stars to navigate for millennia, and in these parts they used, and continue to use, the Southern Cross. It's made up of five stars, and depending on the time of night, it should be easy to find. Look for the Milky Way—the Southern Cross is plopped near a darker spot in the band of our home galaxy.

STATE
**ALASKA**

ANNUAL VISITATION
**598,000**

YEAR ESTABLISHED
**1980**

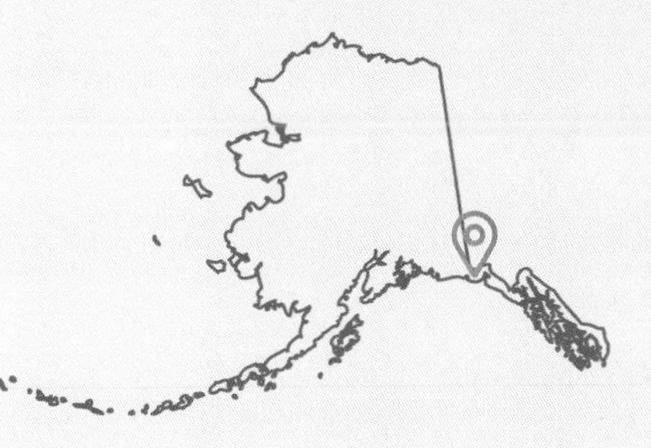

# glacier bay

## coastal mountains & rivers of ice

| | |
|---|---|
| CROWD-PLEASING HIKES | Tlingit Trail (easy); Bartlett River Trail (moderate); Bartlett Lake Trail (strenuous) |
| NOTABLE ANIMALS | Sea otter (*Enhydra lutris*); humpback whale (*Megaptera novaeangliae*); tufted puffin (*Fratercula cirrhata*; see page 316) |
| COMMON PLANTS | Sitka spruce (*Picea sitchensis*); bull kelp (*Nereocystis luetkeana*); chocolate lily (*Fritillaria camschatcensis*) |
| ICONIC EXPERIENCE | Seeing a glacier from the water |
| IT'S WORTH NOTING | You can see a humpback whale skeleton on display near the Bartlett Cove Visitor Center. |
| SISTER PARK | Kenai Fjords (KEFJ) |

**WHAT IS GLACIER BAY?** Glacier Bay is a vast, mountain-bordered bay in southern Alaska. Only a few hundred years ago it was covered by an immense glacier that has since staged one of the fastest glacial retreats ever recorded. GLBA protects not only the bay and its many glaciers, both large and small, but also rugged mountains that rise straight from the sea. Ice and rock aren't the only things on display, though. GLBA's rich waters host an abundance of life sure to make your jaw drop, especially during the brief summers: huge whales, chunky seals hauled out on floating ice, and a wealth of birds rarely witnessed from land.

# WILDCARD

## ICE, ICE, DATA

The force of ice compacting downward traps air bubbles under pressure. If you're lucky enough to get close to a glacier, listen for the popping sound bubbles make as they escape.

**VIEWING A GLACIER IS A** multisensory experience. There are colors: the white and blue of ice, which is also marked with brown and black dirt and debris. There are sounds: the ice groaning, the water splashing, the call of birds overhead, the chatting of your fellow travelers. There is, of course, the feeling of cool sea air and, though hopefully not, perhaps a note of seasickness too. There's a lot going on! But what you can see and experience from onboard a ship or boat (the way most visitors experience this coastal park) is nothing compared to the information stored inside that ice.

Think of glacial ice as one of the world's largest data repositories. The way in which glaciers form—repeated snowfall compacting into ice over long timescales—imprints layer after layer of information. To those who know how to read them, the layers of ice paint vivid pictures of the past and could very well offer a glimpse of the future too.

Take a look at one of the park's glaciers. Its colors are understood to be largely the result of a combination of factors such as age, weathering, ice crystal structure, and air bubbles trapped within the ice. White ice is typically more weathered and contains many bubbles; blue ice is much denser, with fewer bubbles and large ice crystals that scatter light. Deep blue, sometimes turquoise-colored ice comes from the densest and lowest layers of the glacier. It's strikingly saturated hue appears because the dense ice absorbs every other color of light, except blue—so our eyes see it as blue!

The bubbles locked within glaciers do much more than just provide the ice with its mesmerizing colors. Each of those bubbles is like a fossil of Earth's past atmosphere. A layer of fresh snowfall is porous, with lots of air pockets. As the snow begins to compact, the air—and anything in it—gradually becomes entombed in the ice. Each tiny air bubble contains a snapshot of Earth's mix of atmospheric gases (like carbon dioxide and methane) as well as things such as pollen or sulfurous gases from volcanic eruptions swirling in the atmosphere at the time. Sometimes, in the case of massive volcanic eruptions, entire layers of the glacier consist of ash and debris. Glaciers also, of course, pick up rocks and sediments as they move—look for these as dark streaks in the ice. Those trapped sediments are useful indicators of where the ice has been and when it was there.

The glaciers in GLBA tend to be somewhat small and quickly moving—flowing only a short distance from the mountains to the sea. In places where glaciers are larger, cover huge areas, and change less over time across huge distances, scientists extract samples of glacial ice, sometimes from very deep inside the glacier. Known as ice cores, these cylindrical tubes of ancient ice help reconstruct the plants and environmental conditions that shared the world with the ice when it was still fresh snow. Plotting all of it together paints a startlingly detailed picture of climate history on Earth. Knowing the past, of course, is one step toward understanding and preparing for the future. As human-caused climate change advances, having a roadmap for Earth's response to past extreme changes can only help.

Glaciers are beautiful to look at, and their raw power is truly awe-inspiring. Seeing them is a highlight that draws many visitors to GLBA and the other Alaska parks. Their sights and sounds alone are intriguing, but their real power is as storytellers. They are the record keepers of Earth's recent past, a storehouse of information and cautionary tales. Stand in their presence, take in the sights, and relish the opportunity to be in the presence of ice that knows so much.

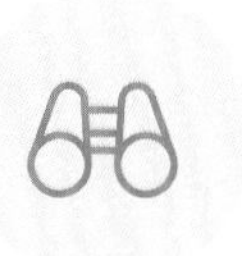

**Several of the park's glaciers meet the sea, so most park visitors see them from the deck of a ship or boat. If you have the time and skill, consider planning a kayak trip to one of the other glaciers not reachable by motorized traffic.**

## HUMPBACK WHALES: OF SCALE AND SMELL

**IN JULY 2001,** a humpback whale was struck and killed by a cruise ship in the park. Her tragic death presented an opportunity and a challenge. Mounting the skeleton for display would let visitors get up close and understand just how big these amazing animals are. But whale skeletons are notoriously oily, which gives them a very powerful smell. Visitors wouldn't love it, but more important, it would problematically attract wildlife. It took more than a decade of off-and-on labor to clean and prepare the bones. Today, the whale, whose Tlingit name is *Tsalxaan Tayee Yáay* (Whale Beneath Mt. Fairweather), can be found near the visitor center in Bartlett Cove.

Adult humpback whales are often the size of a bus.

STATE
**ALASKA**

ANNUAL VISITATION
**595,000**

YEAR ESTABLISHED
**1917**

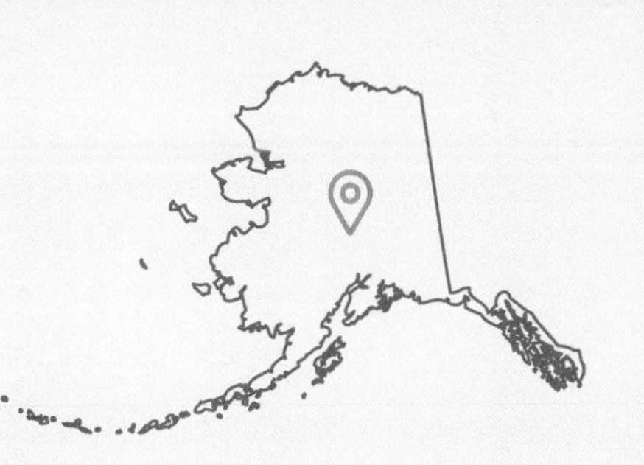

# denali

## rugged peaks & lush forests

| | |
|---|---|
| SUPERLATIVE | Denali (the mountain) is the tallest peak (20,310 feet) in North America |
| CROWD-PLEASING HIKES | Mountain Vista Trail (easy); McKinley Station Trail (moderate); Triple Lakes Trail (strenuous) |
| NOTABLE ANIMALS | Grizzly bear (*Ursus arctos horribilis*); wood frog (*Lithobates sylvaticus*); black-billed magpie (*Pica hudsonia*) |
| COMMON PLANTS | Black spruce (*Picea mariana*); bog blueberry (*Vaccinium uliginosum*); Bigelow's sedge (*Carex bigelowii*) |
| ICONIC EXPERIENCE | Hoping that the clouds will clear to reveal Denali's peak |
| IT'S WORTH NOTING | Denali was the first national park set aside to protect wildlife. |

**WHAT IS DENALI?** Named for and home to the highest peak in North America, DENA is the rugged heart of interior Alaska. It's known for its abundant wildlife, so you'll likely see majestic mammals and soaring birds while you're taking in the scenery. Since Denali is such a massive mountain, it actually creates its own weather, so you may miss out on seeing the peak; it's often shrouded in clouds. Fear not, though, because there's more to the park than the peak. Enjoy hikes near rivers during the summer; and if you brave an Alaskan winter, you'll see a night sky you'll never forget.

# WILDLIFE

## FROZEN FROGS

Male wood frogs identify females only with touch—they basically hug other frogs to see if they're plump. If they're not plump, it means that they're males or they're females that have already laid their eggs.

**DENA IS ONE PARK WHERE** you're sure to see some incredible wildlife. From the formidable grizzlies to the iconic cold-loving caribou, there's no shortage of mammalian life here. With its northern locale, it's not surprising that DENA doesn't have any reptiles, and you'd think it would be lacking amphibians too. Yet there is one amphibian who's decided that cold is just an opportunity for a nice long sleep.

The wood frog (*Lithobates sylvaticus*) is relatively widespread in North America and very common in DENA. They're relatively tiny, often less than 2 inches long. Like their relatives, canyon tree frogs (see page 107), they have an incredibly loud and recognizable song, an indicator that spring has arrived in the park. Males make a loud "racket, racket" or chuckle noise that sounds kind of like a chicken.

What are these frogs waking up from in the spring, anyway? Impressively, they aren't waking up so much as they're *thawing out*. Wood frogs are specially adapted to survive the long, harsh, cold winters in Alaska. The adaptation is a very efficient production of cryoprotectants, which makes it totally cool (get it?) for the frog to freeze and thaw like it's no big deal. Other creatures, including insects, use cryoprotectants to survive frozen conditions. For wood frogs, the key player is glucose. During the short autumn in Alaska, wood frogs get the signal that the freeze is coming, which seems to trigger a change in their livers. They start to convert stored glycogen into glucose, which courses through their blood. The glucose helps protect the frogs from dehydration and other cell damage. They burrow into decaying leaf litter, where their eyeballs and limbs freeze first, and their essential organs freeze last. Unsurprisingly, the frogs are completely lifeless during their deep freeze. Then, come spring, they thaw, the

A thawing wood frog

glucose converts back to glycogen, and the frogs are ready to find a mate and start the cycle all over again.

It's worth noting that these Alaskan frogs thrive with the help of more urea production than the warmer-weather wood frogs of the south. Urea helps keep glucose in the cells prior to freezing, so having a high concentration of it gives the frogs an edge to survive the incredibly harsh conditions. They also have glycolipids, which seem to prevent ice from forming inside the frogs' cells.

While you're traveling through DENA enjoying the diverse wildlife and scenery, keep your ears alert. As you hear the song of the wood frogs, just remember that not too long ago, these frisky fellas were totally frozen, and the first thing they do when they thaw is hoot and howl looking for a mate. The promise of propagation is great motivation.

**You might get lucky and see a tiny wood frog in ponds in forested areas of the park. If not, though, you're sure to hear them during the warmer months!**

## WILDCARD

### GO HOME, FOREST, YOU'RE DRUNK

**IF YOU HAVEN'T BEEN TO** Alaska, there are some things that probably come to mind when you think of this massive state. Untouched wilderness and cold weather might be on your list, and for good reason—Alaska is the northernmost state of the United States and is indeed vast, with a lot of remote areas minimally populated by humans (or not populated at all). During your time in DENA, you can enjoy a slice of the solitude as you walk among the trees and observe the wildlife. If you have a keen eye, you might notice that in the forests, while there are a lot of trees, there aren't a lot of *different* trees—only eight species in the entire park. But these trees have some wisdom to share about cold-weather adaptations and plans for continuing to live in our warming climate.

Among the tree species in the park, there are two spruces that you're likely to see: black spruce (*Picea mariana*) and white spruce (*Picea glauca*). These conifers can look similar, although they're different species. Often, black spruce will be more dwarfed and skinnier than white spruce. As our climate continues to change, scientists have their eyes on melting permafrost, a layer of long-frozen soil, and what it will mean for historically cold landscapes.

Permafrost melting undoubtedly impacts land where the ground has been frozen since time immemorial. Where humans have made homes and roads on permafrost, the landscape shifts when that layer melts, and those structures are damaged or destroyed. When permafrost thaws, pockets of soil thaw preferentially, which causes the ground to shift and sometimes creates deep pits. In DENA, you can see evidence of permafrost thawing by way of a phenomenon casually called "drunken forests." Often dominated by black spruce, tree stands here aren't always upright, and the trees look like they're not able to stand up straight. No, the trees haven't had a rough night out at the bar—they're lopsided because of melting and shifting permafrost. Sometimes,

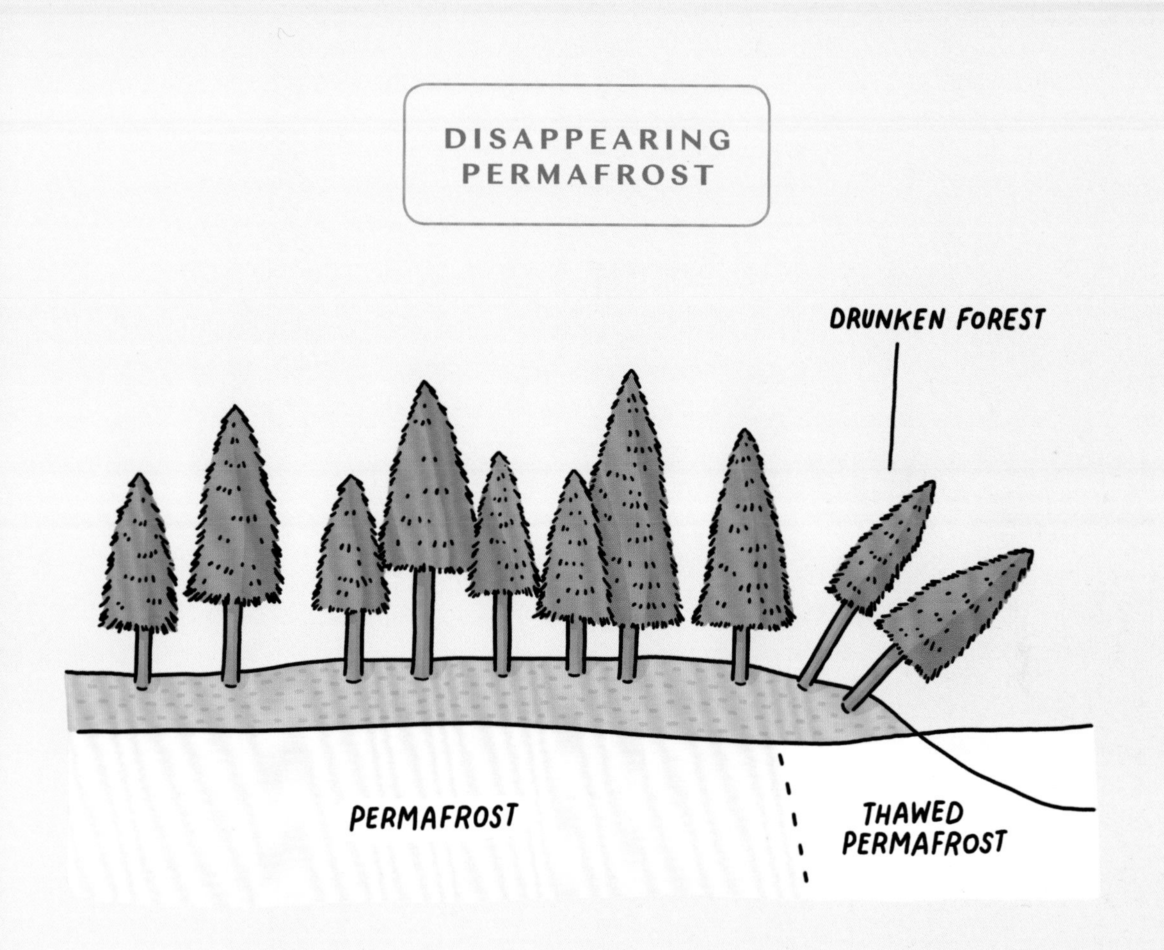


trees can become so displaced by these shifts that they topple over and die.

Understandably, permafrost warming has massive ecological implications for our planet, particularly for the northern parts of the continent, including Alaska. Research on the spruce trees in the park and how climate affects them indicates that there may not be a straightforward way to predict how thawing permafrost will affect the region's forests. Apparently climate isn't the only factor that affects spruce success in DENA. Other factors impact individual stands, including elevation and slope angles. It's possible that as climate changes and permafrost thaws, stands dominated by black spruce might give way to white spruce, allowing a fertile forest to continue to grow. This is not to say that thawing permafrost should be ignored, rather that the forest might continue to find a way as the climate warms.

As you see spruces in different stages of sobriety, don't forget to consider the ground beneath them. Our planet has undoubtedly been impacted by the actions of humans, and DENA is one place where we can see some changes happening early on. However, life here has always found a way, and we may be cheered to know that these forests won't go down without a fight.

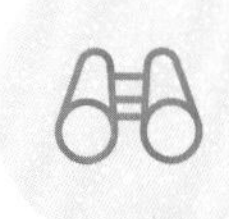

**Keep your eyes peeled throughout the park for trees unable to stand up straight. One place to see them is just north of Sanctuary River Campground, where you can see black spruce trees in various positions due to melting, shifting permafrost.**

STATE
**ALASKA**

ANNUAL VISITATION
**322,000**

YEAR ESTABLISHED
**1980**

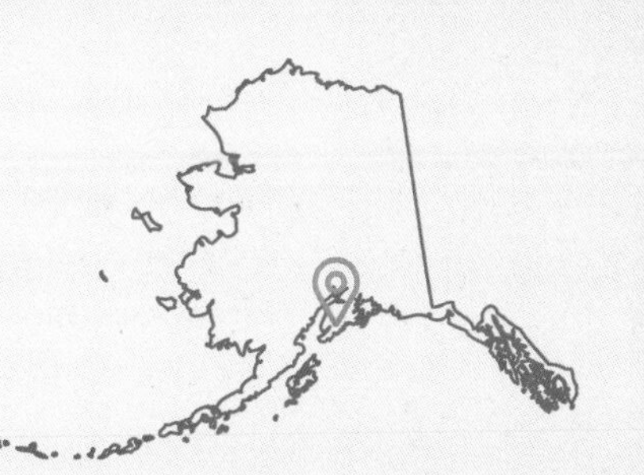

# kenai fjords

## mountains & glaciers in a coastal wonderland

| | |
|---|---|
| SUPERLATIVE | KEFJ is the smallest national park in Alaska |
| CROWD-PLEASING HIKES | Glacier View Loop Trail (easy); Glacier Overlook Trail (moderate); Harding Icefield Trail (strenuous) |
| NOTABLE ANIMALS | Mountain goat (*Oreamnos americanus*); black bear (*Ursus americanus*; see page 252); bald eagle (*Haliaeetus leucocephalus*) |
| COMMON PLANTS | Dwarf fireweed (*Chamaenerion latifolium*); Sitka alder (*Alnus crispa*); mountain hemlock (*Tsuga mertensiana*) |
| ICONIC EXPERIENCE | Exploring the park's waters on a boat tour |
| IT'S WORTH NOTING | The park is open year-round and is a great place for snowy activities such as cross-country skiing and snowshoeing. |
| SISTER PARK | Glacier Bay (GLBA) |

**WHAT ARE KENAI FJORDS?** No one knows where the name *Kenai* came from (it's possibly a corruption of an Indigenous or Russian word), but fjords—narrow ocean inlets surrounded by steep cliffs—are abundant in this coastal park. Home to dozens of glaciers and hundreds of miles of coastline, KEFJ gives visitors the chance to make memories on land, in the air, and on the water. Hike to a glacier to witness its retreat, board a flightseeing plane to explore a massive ice field, or hit the water in a tour boat or kayak to get a glimpse of seabirds and whales. Whatever the season, your KEFJ experience is one you won't soon forget.

## WILDLIFE

### THE NOMAD'S LIFE FOR ME

Puffins shed their colorful beak plates once a year, revealing a much duller, gray beak.

**ALASKA'S PARKS ARE SOME OF** the wildest places on our planet. Bears, mountain goats, hares, and lynx are just a few of the mammals comfortable in the cool temperatures of our northernmost state. It's easy to get distracted by our abundant mammalian cousins, and to have them at the top of your sightseeing list. However, in KEFJ, you have a chance to see some of the world's most famous seabirds—puffins! Two species breed in the park, and you just might see them if you head out on the water.

There are four species of puffin in the world, and the two that can be found in KEFJ are the tufted puffin (*Fratercula cirrhata*) and the horned puffin (*Fratercula corniculata*). They're really similar in size, 15 to 16 inches tall, about the size of a bowling pin. Both have the distinctive colorful feet and beaks for which puffins are known, along with white feathers on their faces. Aptly named, you can distinguish a tufted puffin by the yellow feathers atop its head, while horned puffins have a fleshy black horn above each eye.

Puffins are nomads; they spend the vast majority of their lives at sea. It makes sense, then, that they're much stronger swimmers than they are flyers. When you see them in the air, notice their frantic-looking flapping. They'll typically flap up to 400 times per minute, which makes them look much weaker in the air compared to something that glides confidently about, like a red-tailed hawk (see page 170).

The only time these seabirds return to land is when they come to the islands and surrounding waters of KEFJ in May to breed and raise their young. Both species meet as a large group in the waters along the shore at slightly different times, with tufted puffins arriving first. Since both species mate for life, one theory is that this group meeting on the water is a chance for the mates to reunite. When they move onto land, the pairs build their nests and each care for one egg. Horned puffins make their nests in rocky crevices; tufted puffins dig deep burrows. They feed on fish in the

## JUST PUFFIN AROUND

Even though tufted puffins and horned puffins look similar, you can tell them apart because of the tufted puffins' yellow head feathers, and the black, fleshy horns above the eyes on the horned puffins.

Tufted puffin
(*Fratercula cirrhata*)

Horned puffin
(*Fratercula corniculata*)

When fledgling puffins of both species leave their nests, they live at sea for two to three years before they come back to find a life mate.

nearby waters, and the reliable food here may be what draws them back every year. They typically head back to sea in September to live their nomadic seabird life once more.

Not surprisingly, climate change is a cause for concern for these nomadic birds. As Alaska's waters warm, fish and other aquatic wildlife struggle to thrive. Since it's a delicate, interdependent food web, when smaller animals founder, larger predators struggle too. While you're in KEFJ enjoying the wildlife, glaciers, and stunning waters, consider that the life you see on land is intimately connected with life in the water and will need to find a way to continue to thrive as human-caused climate change shifts the Alaskan ecosystem.

**Both species of puffin can typically be found in several areas of the park by way of boat tours between May and September. Both Beehive Islands (so named for the buzzing of puffins going out to sea for food and back) have large populations of puffins. Other good spots include Emerald Cove (horned puffins) and just beyond Cheval Island (tufted puffins).**

STATE
ALASKA

ANNUAL VISITATION
79,000

YEAR ESTABLISHED
1980

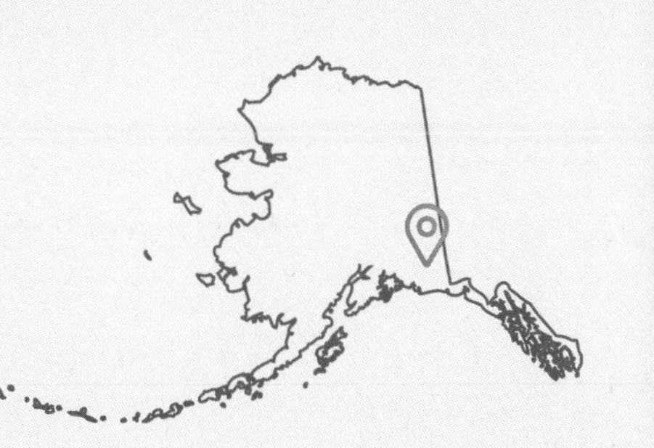

# Wrangell-St. Elias

## unimaginable scale in splendid isolation

| | |
|---|---|
| SUPERLATIVE | WRST is the biggest park in all the land—six times larger than YELL |
| CROWD-PLEASING HIKES | Toe of the Kennicott Glacier Trail (easy); Skookum Volcano Trail (moderate); Bonanza Mine (strenuous) |
| NOTABLE ANIMALS | Red squirrel (*Tamiasciurus hudsonicus*); hairy woodpecker (*Picoides villosus*); moose (*Alces alces*) |
| COMMON PLANTS | Forget-me-not (*Myosotis asiatica*); blueberry (*Vaccinium* spp.); black spruce (*Picea mariana*; see page 312) |
| ICONIC EXPERIENCE | Staring at a map and then a mountain range (or two or three or four . . .) in disbelief at just how big and remote this park is |
| IT'S WORTH NOTING | Flightseeing (touring via small plane) is a great way to see otherwise inaccessible parts of this huge park. |

**WHAT IS WRANGELL-ST. ELIAS?** The Wrangells and the St. Elias are two of the four (!) mountain ranges within WRST. (Also included, since you asked, are the Chugach Range and the Alaska Range.) This enormous park boggles the mind. It contains nine of the sixteen highest peaks in the United States and a dazzling collection of enormous glaciers, one of which is larger than Rhode Island. A historic mining town sits in contrast to a wild land that has sustained more than 10,000 years of human history. Mountains after mountains filled with a rich and wild diversity of plants and animals await anyone who ventures into the edge of this wilderness.

# BOTANY

## SUMMER'S TIMEKEEPER

Many parts of fireweed are edible. While you're in Alaska, keep your eyes peeled for syrups, jams, teas, salads, and side dishes made from the plant.

**TIME WORKS A LITTLE DIFFERENTLY** in Alaska, especially during the long, light-filled summer. In a place that so easily dwarfs your sense of scale, it can be easy to lose your mooring to time as well. The good news is that a humble and beautiful flower is here to help you keep track of summer's progress and remind you that destruction is often a precursor to color and life.

Fireweed (*Chamerion angustifolium*) is one of the most common flowers in Alaska, and WRST is no exception. During the summers, you won't have to look hard to find the tall stalks of pink-purple blooms. They paint the sides of park roadways, decorate the front yards of McCarthy residents, pop up along streams and riverbanks, burst forth in fields in front of glaciers, and show up in just about any open sunny space from sea level to alpine zone. Fireweed thrives in full sunlight and excels at growing on disturbed sites—in fact, its name comes from its propensity for taking over recently burned land. You will undoubtedly snap a plethora of pictures during your time in WRST, and odds are good that you will see fireweed in many of them.

The only time you'll need to look hard to find fireweed is at the very beginning of the season, soon after the snow melts. The new shoots look like a strange dark-red asparagus—they are remarkably well camouflaged against the leaf litter and twigs they push through. As summer progresses, fireweed trades in relative anonymity for a full-throated proclamation. Growing quickly, the unbranched stems are soon 4 to 6 feet tall.

By July, the top quarter or so of the stem will be covered in fleshy buds. Starting from the bottom and working upward, those buds burst forth into brightly colored, four-petaled flowers. Keep an eye out for bees, butterflies, and hummingbirds hovering and pollinating nearby. After the flowers have bloomed themselves out, they fade and are replaced by thin, slender pods that start curling backward as they release their seeds. Fireweed blooms slowly and methodically over the summer months—pay attention and you'll quickly spot plants that have seed pods on the bottom, open flowers in the middle, and unopened buds on top.

By the time all the flowers have faded and the pods have opened to release thousands of downy white seeds, the entire plant turns a flame red. Continually shifting throughout the season, fireweed's colors are a marker of summer and a harbinger of change. Traditional Alaskan knowledge holds that the first snows will arrive about six weeks after the fireweed fades. Note that in WRST, as in just about all mountainous places, going up in elevation is like going back in time. If you miss the fireweed blooms at lower elevations, simply head upward, where cooler temperatures stave off blooms until later in the season.

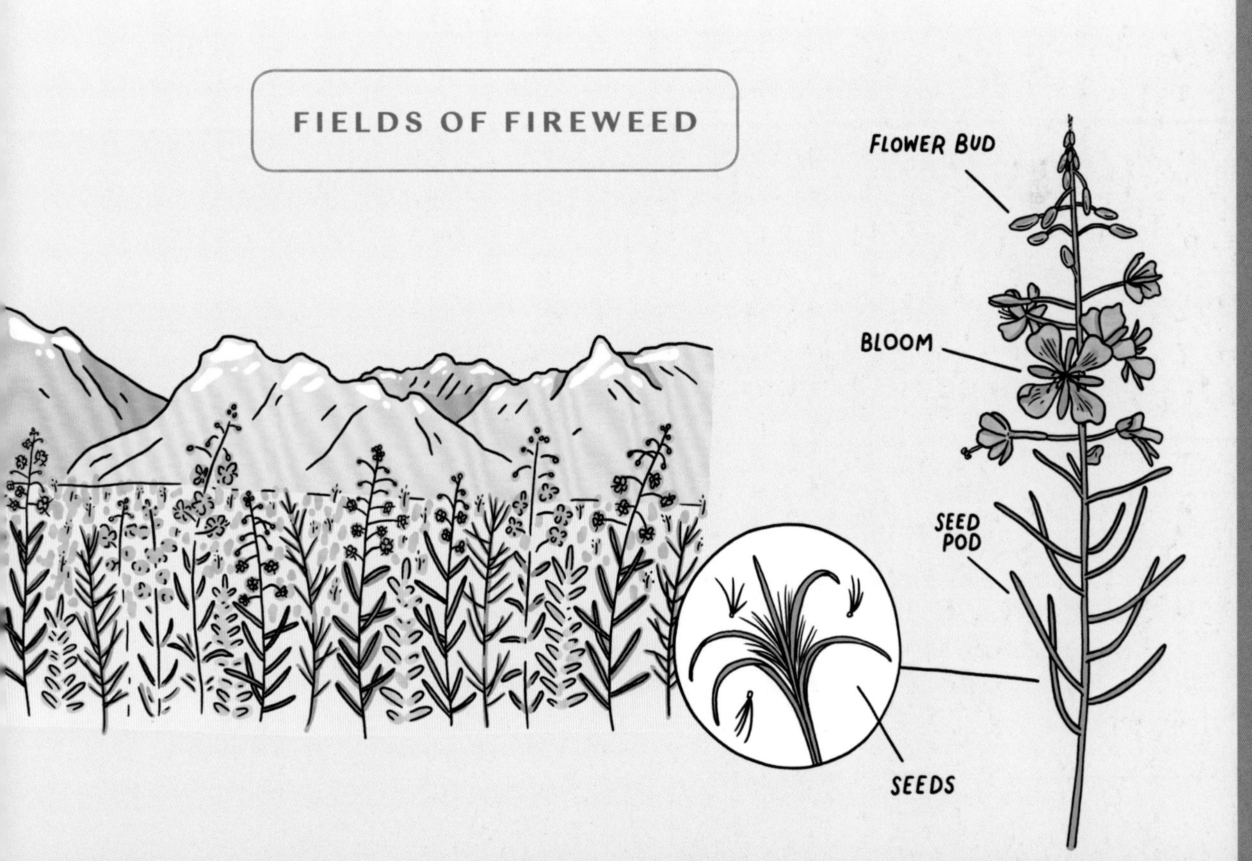

Fireweed reproduces both by seeds, which are readily distributed by the wind, and by sprouting, through its extensive network of underground rhizomes.

Fireweed is native to most of North America, where it is often closely associated with recovery from natural disturbances such as volcanic eruptions and forest fires. It holds those same associations here in Alaska, but in a place where summers are short and warm weather is fleeting, the carpets of pink flowers and blankets of white, fluffy seeds take on a special significance. Fireweed is a botanic calendar, a bright and beautiful reminder that even after a long winter, summer is ours to soak up the sun and be as brilliant as we'd like.

**Fireweed is common throughout the park. It typically blooms from June through August.**

STATE
**ALASKA**

ANNUAL VISITATION
**38,000**

YEAR ESTABLISHED
**1980**

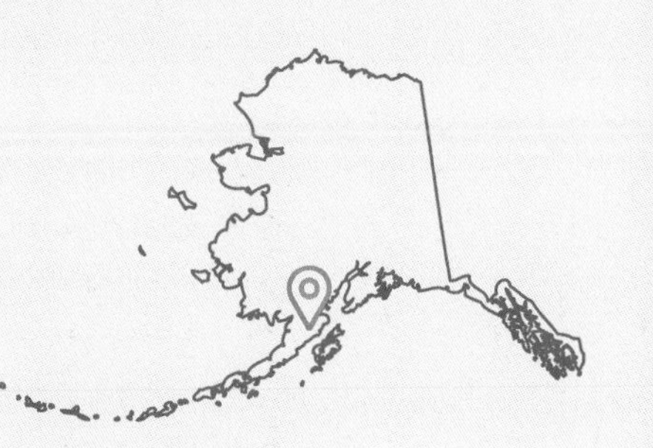

# Katmai

## a volcanic wonderland, alaska-style

| | |
|---|---|
| SUPERLATIVE | KATM is the site of the largest volcanic eruption of the twentieth century |
| CROWD-PLEASING HIKES | Brooks Falls Trails (easy); Dumpling Mountain Trail (moderate); Novarupta (strenuous) |
| NOTABLE ANIMALS | Brown bear (*Ursus arctos*); sockeye salmon (*Oncorhynchus nerka*); common goldeneye duck (*Bucephala clangula*) |
| COMMON PLANTS | Fireweed (*Chamerion angustifolium*; see page 320); Nootka lupine (*Lupinus nootkatensis*); common yarrow (*Achillea millefolium*) |
| ICONIC EXPERIENCE | Watching bears catch salmon on the Brooks River |
| IT'S WORTH NOTING | Most visitors see only a tiny portion of this immense park, which is larger than YELL and YOSE combined. |

**WHAT IS KATMAI?** Mount Katmai is one of no fewer than a dozen volcanoes within the boundaries of KATM. Fresh evidence of volcanism is everywhere—especially in the park's famous Valley of Ten Thousand Smokes, a fresh scar from the massive eruption that rocked this area in 1912. The park is located on the Alaskan Peninsula, the northern rim of the Pacific Ring of Fire. Visitors come to get up close with the park's famed wildlife, such as huge coastal brown bears plucking salmon out of the Brooks River. Wild creatures amid an even wilder landscape make visiting KATM a rugged and unforgettable experience.

# GEOLOGY

## LIGHT-WEIGHT TALES OF HEAT AND POWER

Leave the pumice where you found it, please! Removing it is illegal.

**IN 1912, A NEW VOLCANO** in the Katmai volcanic center announced its arrival on the scene with a bang. Over the course of three days, Novarupta spewed more magma than any other volcano in the twentieth century. Novarupta's eruption was so powerful that it drained magma from under Mount Katmai, 6 miles away, causing Katmai to collapse into a caldera more than 2 miles wide and ½ mile deep. When it was all over, the eruption had covered a lush valley in hundreds of feet of volcanic debris and unleashed thirty times more magma than Mount St. Helens would produce later in the century.

You might not expect such a powerful eruption to produce rocks light enough to float, but that is exactly what happened. The magma under this part of Alaska is thick like tar; it does not flow smoothly like the magma fueling the Hawaiian hot spot (see page 288). Instead, pressure builds underground until it explodes. In 1912, the first explosion sent a column of debris more than 18 miles into the sky. When gas-rich magma suddenly depressurizes as it's thrown from the volcano, it shatters into fragments that have different names based on their size. Though large blocks sometimes fall, most often this initial eruptive phase produces ash (very fine, powdery particles) and pumice (fist-size chunks of extremely porous rock).

Looking around KATM today, whether from your base at Brooks Camp or as you explore the Valley of Ten Thousand Smokes, you're likely to see an abundance of pumice. These air pocket–filled rocks are so light they float on water—a classic sight along the beaches of Nanek Lake—but look closely: Pumice has stories to tell. The 1912 eruption, in which Novarupta spewed an incredible amount of magma, was somewhat brief but incredibly complex. The chemical composition of the magma changed over the course of the three-day eruption, producing different types of pumice that you can still see in many parts of KATM. Pay attention, and soon you'll be reading pumice like a pro.

## PEEPING PUMICE

Use this handy chart to identify the pumice you find around KATM.

The main difference between types of pumice is how long the rocks were stored underground before being thrown skyward and how much silica they contain. The most silica-rich pumice produced by Novarupta is a nearly uniformly white rock called *rhyolite*. The lack of visible mineral crystals (which would appear as specks of color) indicates that the rhyolite didn't do much of its cooling underground. The high silica content makes rhyolite thick like tar. Gases trapped inside the magma struggle to escape, making rhyolite the most explosive type of magma. If you spot a white or gray rock that's covered in darker specks, it's likely to be *dacite*—kind of the middle-of-the-road pumice. It isn't as thick or explosive as rhyolite and has spent more time underground, thus forming

Occasionally, you may come across brightly colored pumice that might be yellow, red, or orange. That means the pumice was altered by extreme heat after it was erupted.

those nice mineral crystals (specks) for you to see. Brown to black pieces of pumice, usually decorated with light-colored specks, are *andesite*—a silica-poor magma that spent enough time underground to grow some visible crystals and is much less explosive than rhyolite. Keep an eye out for pumice that is striped and banded in different colors, kind of like marble cake. It looks like different types of magma swirled around each other—because that's what it is! Though scientists had long speculated that different types of magma might mix and mingle underground, these rocks were some of the first hard evidence for such a phenomenon.

Rhyolite, dacite, and andesite were ejected together during the eruption, but the relative proportions varied over time. These variations allow scientists to reconstruct the eruption timeline and understand more about how, when, and where Novarupta's volcanic debris spread through this area and ultimately around the world.

Today, as you move through KATM, keep an eye out for pumice. The fact that it's a floating rock is really just a neat party trick when you stop to consider the larger story. That porous, lightweight rock is a piece of the inner earth that has been shot miles into the sky while being blasted apart by exploding air bubbles. It fell back down, where its new job is to serve as a postcard from underground, telling tales of heat, power, and time, cleverly disguised as subtle colors.

**Pumice is abundant throughout large parts of KATM, especially along the waterfront of Brooks Camp and the Valley of Ten Thousand Smokes itself.**

STATE
**ALASKA**

ANNUAL VISITATION
**15,000**

YEAR ESTABLISHED
**1980**

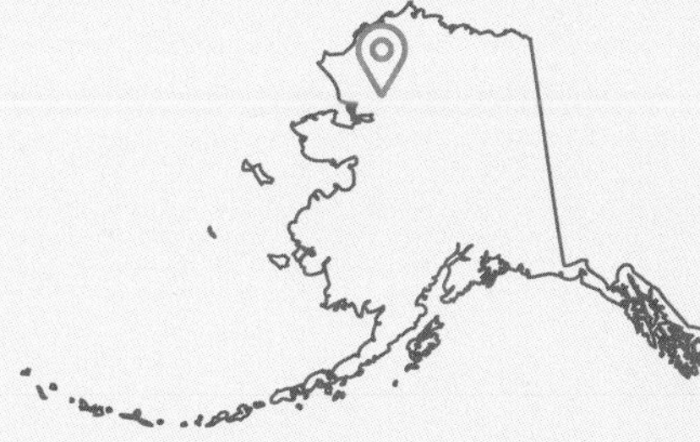

# kobuk valley

## immense dunes & migrating mammals

| | |
|---|---|
| SUPERLATIVE | KOVA is home to the largest active sand dunes in the Arctic |
| CROWD-PLEASING HIKES | There are no public roads, entrance stations, established trails, or facilities in the park. |
| NOTABLE ANIMALS | Caribou (*Rangifer tarandus*); wood frog (*Lithobates sylvaticus*; see page 310); chum salmon (*Oncorhynchus keta*) |
| COMMON PLANTS | Kobuk locoweed (*Oxytropis kobukensis*); black spruce (*Picea mariana*; see page 312); paper birch (*Betula papyrifera*) |
| ICONIC EXPERIENCE | Watching the two hundred thousand or more caribou during their annual migration |
| IT'S WORTH NOTING | Onion Portage has been used as a prime caribou hunting site for over 8,000 years, and the Inupiat people still hunt there today. |

**WHAT IS KOBUK VALLEY?** Wild and remote, the Kobuk Valley is a river valley encircled by mountains and steeped in thousands of years of human history. KOVA protects the valley and its river, a wild stretch of land that is much more heavily utilized by local subsistence hunters than by typical park crowds. One of two parks located above the Arctic Circle, it's known for its sand dunes—remnants of glacial advance and retreat over the last 30,000 years or so. Adventurers enjoy rafting the Kobuk River, hiking the dunes, and watching the wildlife that dominate the land. A trip to KOVA is a reminder that there are still places in our country where most humans are mere visitors.

# lake clark

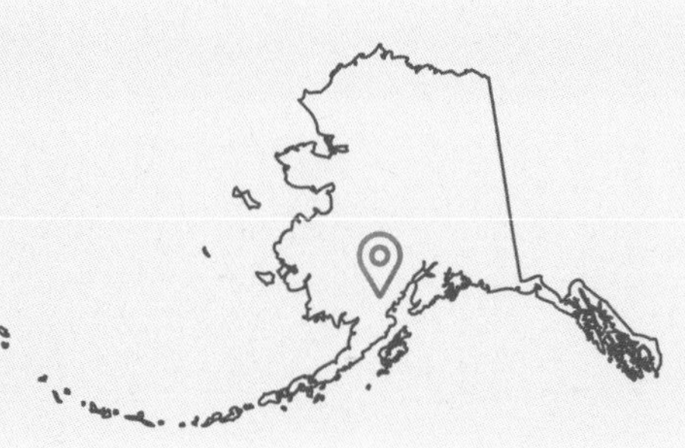

STATE
**ALASKA**

ANNUAL VISITATION
**14,000**

YEAR ESTABLISHED
**1980**

## waterfalls & lakes rimmed by rugged mountains

| | |
|---|---|
| **SUPERLATIVE** | Iliamna Lake is the largest spawning grounds for sockeye salmon in North America |
| **CROWD-PLEASING HIKES** | Tanalian Trails is the only maintained trail system in the park. There are abundant hiking opportunities without maintained trails as well! |
| **NOTABLE ANIMALS** | Brown bear (*Ursus arctos*); moose (*Alces alces*); sockeye salmon (*Oncorhynchus nerka*) |
| **COMMON PLANTS** | Fireweed (*Chamerion angustifolium*; see page 320); Ramensk's sedge (*Carex ramenskii*); salmonberry (*Rubus spectabilis*) |
| **ICONIC EXPERIENCE** | Landing on the water in a float plane! |
| **IT'S WORTH NOTING** | The Dena'ina have lived here for over 9,000 years and continue to live off the land today. |

**WHAT IS LAKE CLARK?** Lake Clark is one of the largest lakes in Alaska and just one of the many stunning freshwater bodies that dot LACL. The park protects the headwaters and spawning grounds for one the most abundant salmon runs in North America. Located on the Pacific Ring of Fire, LACL is also a geologic wonderland featuring mountains, two active volcanoes, and numerous glaciers. It's also a coastal park, with beaches and salt marshes that provide unmatched opportunity to observe brown bears along with other wildlife such as wolves. Whether you explore the park on foot, by plane, or in a kayak, you're sure to make memories while you breathe the clear Alaskan air.

STATE
**ALASKA**

ANNUAL VISITATION
**10,000**

YEAR ESTABLISHED
**1980**

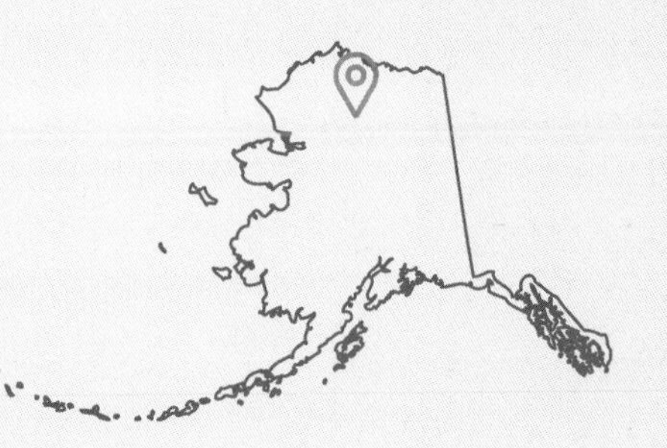

# gates of the arctic

## remote wilderness north of the arctic circle

| | |
|---|---|
| SUPERLATIVE | The mountains here are the northernmost range of the Rockies |
| CROWD-PLEASING HIKES | There are no established trails in the park |
| NOTABLE ANIMALS | Wood frog (*Lithobates sylvaticus*; see page 310); snowshoe hare (*Lepus americanus*); common loon (*Gavia immer*) |
| COMMON PLANTS | Black spruce (*Picea mariana*; see page 312); bear flower (*Boykinia richardsonii*); arctic lupine (*Lupinus arcticus*) |
| ICONIC EXPERIENCE | Taking a flightseeing trip to see the park from the sky |
| IT'S WORTH NOTING | This is remote and challenging terrain; visitors who wish to explore on the ground should be trained in survival skills and come prepared to fend for themselves. |

**WHAT IS GATES OF THE ARCTIC?** Named for its position near the far northern Arctic, GAAR provides some of the best opportunities for solitude and wildlife viewing anywhere on the continent. It's the second-largest national park in the system, and no roads or trails cross this vast wilderness. People have occupied this land for over 10,000 years, and there is an Inupiat village within the park boundaries today. A backpacking trip here lets the imagination go wild about what it would have been like to visit thousands of years ago. Six rivers run through the park, along with other waterways that typically thaw in June, when life briefly returns to the otherwise frozen land.

## WILDCARD

### SEEING THE DARK

**FOR MANY PEOPLE, ALASKA REPRESENTS** something like the great unknown: It's cold, the environment is harsh, and the sights are guaranteed to be different than your day-to-day life. It's a place that can help recalibrate your sense of wonder, with its huge mountains, sprawling glaciers, churning seas, and sparkling lakes. The remote location and stormy climate make Alaska challenging to get to. But once you've gone all the way there, why not go one step further? Grab some warm clothes and see if you can't hitch a ride to space.

Of course, you can't *actually* board a rocket ship in Alaska—launching from near the tropics is a much better call—but the night skies in GAAR, KOVA, and LACL are stunning enough to give you the sensation of having left the planet, all while your wool sock–clad feet stay firmly on the ground. On clear, dark-sky nights, all three parks will afford you the opportunity to stand under a Milky Way that stretches from horizon to horizon. Even if you've seen the Milky Way in other places or parks, seeing it here can be a whole new experience. That's because the remote locations of these three Alaskan parks make them some of the only places left in the United States where human-created light pollution is virtually nonexistent. That lack of anthropogenic light means it's sometimes possible to see incredible details and phenomena that are obscured even in places where the night sky is otherwise brilliant.

While your eye is naturally drawn to the twinkling lights of the Milky Way, don't neglect its darkness! Running through the middle of the Milky Way is a dark band known as the Great Rift or the Dark Rift. In reality, the rift isn't empty or dark. It's a huge region of dust clouds sitting between our solar system and the next closest arm in our spiral-shaped galaxy, blocking our view of the stars beyond. Within the dust clouds (not visible to us because the dense dust does not allow visible light to shine through) are stars in the very early stages of their formation. The Great Rift is an enormous stellar nursery!

The extremely dark skies in these parks also mean you may be able to catch a glimpse of the phenomenon known as zodiacal light or false dawn. This eerie glow looks like a hazy pyramid that appears in the sky just before true dawn in the months around the autumn equinox. Seeing this strange light is awe-inspiring on its own, but even more so when you know its origin. The light you're seeing is sunlight glinting off dust grains in the inner solar system. Those grains, by the way, are leftovers from when the rocky planets of our solar system first took shape about 4.5 billion years ago.

Earth is not the only planet in our solar system (or beyond) with auroras. Any planet that has a magnetic field is likely to experience some aurora activity. Stunning displays on Jupiter and Saturn have already been observed by space probes.

Being so far north means that not only do you have the opportunity to see the Milky Way, you also stand an above-average chance of seeing the dancing curtains of the northern lights during winter and spring. That is especially true for KOVA and GAAR, both of which are situated above the Arctic Circle. More properly called the *aurora borealis*, the northern lights are caused by charged particles from the sun slamming into Earth's upper atmosphere. (From the Southern Hemisphere, you can see the *aurora australis*, or southern lights.) Solar wind whisks particles away from the sun and toward the rest of the solar system along magnetic field lines. The particles interact with gases in our atmosphere, producing vivid displays of colors. Different molecules produce different colors; oxygen most commonly emits green and red light, though sometimes it can skew yellow. Nitrogen is responsible for blue and purple hues. Think of the lights as a postcard from the sun: *Hey there. Been windy here lately. Sending lots of particles your way. My love to your magnetic field. XOXO, The Sun*

The stunning sights and extreme isolation of the three least-visited parks in the system are thrilling experiences all on their own. But time your visit right and you could be treated to not just views of rugged terrain but a view that stretches out toward the edges of our solar system, our whole galaxy, and beyond.

**Alaska is well known for its long, light-filled summer days. To experience the aurora and zodiacal light, you'll need to travel during the spring or winter and plan to stay up late. Challenging, but worth it! The "northern lights" season in Alaska lasts from approximately mid-September to late April; activity peaks in March.**

# good books

You can find a full list of the sources we consulted when writing this book at www.scenicsciencebook.com. Here is the short list of books that we found exceptionally helpful. (And in many cases, the authors of these books were personally helpful to us as well—thanks, y'all!)

*The Bristlecone Book: A Natural History of the World's Oldest Trees*, Robert M. Lanner

*An Ecotourist's Guide to the Everglades & the Florida Keys*, Robert Silk

*Fireflies, Glow-worms, and Lightning Bugs*, Lynn Frierson Faust

*Geology along Skyline Drive: A Self-Guided Tour for Motorists*, Robert L. Badger

*Great Smoky Mountains National Park: A Natural History Guide*, Rose Houk

*The Little Book of Sea-Beans and Other Beach Treasures*, Cathie Katz and Paul Mikkelsen

*Roadside Geology* series, various authors

*Stars Above, Earth Below: A Guide to Astronomy in the National Parks*, Tyler Nordgren

*The Southwest Inside Out: An Illustrated Guide to the Land and Its History*, Thomas Wiewandt and Maureen Wilks

*Tidepools of the Pacific Coast: A Quick Field Guide*, Michael Rigsby

*The White River Badlands: Geology and Paleontology,* Rachel C. Benton, Dennis O. Terry, Jr., et al.

# about the authors

EMILY HOFF

**EMILY HOFF** is the Big Ideas Person of this author duo. She's constantly inventing new projects and swinging for the fences. Other hats that Emily wears when working with Maygen include Chief Campsite Dishwashing Officer, Monitor of the Weather Radar, Knower of *Hamilton* Lyrics, and Queen of the Research Books. When she isn't working with Maygen, you can find her writing museum exhibits for the likes of the American Museum of Natural History and Kennedy Space Center. She lives in Brooklyn, New York.

MAYGEN KELLER

**MAYGEN KELLER** is the Pragmatist in Chief of this author duo. Her organization, wit, and budgetary skills are constantly being called upon to make Big Ideas into Reality. Known for her ability to quickly assemble tents, she's also renowned throughout the country for being able to start a campfire in a rainstorm. When working with Emily, her roles also include Chief Car Organization Architect, Chief Playlist Officer, Lead Harmonizer in *Hamilton* Sing-Alongs, and #Hashtag Genius. In other parts of her life, she is a creative project manager and salesperson. She lives in Minneapolis, Minnesota.

# about the illustrator

**JILLIAN BARTHOLD** is an illustrator based in Portland, Oregon, who once visited twelve national parks in thirty days. She can usually be found drawing plants and small treasures. See more of her work at jillianbarthold.com.

# acknowledgments

So many people touched this project over the years and we owe a great deal to all of them. A few deserve very special thanks. Dennis Terry and Tyler Nordgren crazily answered our calls from the very beginning. They have been guiding (star)lights and sources of inspiration over the life of this project. Likewise, Amanda Yates and Diana Montano poured their wonderfully creative professional selves into what was at the time nothing more than a crazy idea. Chuck Schroll remains the only pilot and novice whiskey drinker for us. Kaitlin Ketchum, our incredible editor, and the rest of the Ten Speed Press team, including Lizzie Allen, Doug Ogan, Dan Myers, Andrea Portanova, and Felix Cruz, made this book a reality. Jillian Barthold's whimsical illustrations brought our words to life in a way we never imagined possible. And finally, our amazingly gifted and truly hard-working agent, Cindy Uh, believed in us and this book from Day One. There's no doubt that this book's publication is largely a credit to her.

We would also like to say a hearty THANK YOU to the friends and family who fed us, housed us, got us a hotel room here and there, made sure we had the right gear, lent us the love of their pets, and occasionally poured us a drink and wiped away the tears. We love you! Special thanks to Ethan Angelica for unlimited living room use, and to the Hoff-Carlson family who let us turn their quiet home into our Western HQ for two summers.

The following experts in their fields, many from within the National Park Service, checked parts of this book for accuracy and clarity. We are extraordinarily grateful for their helpful and enthusiastic responses. Every single one of them made this book stronger; any errors are our own. **OLYM**: Dean Butterworth, Janet Coles, Steve Fradkin, Jared Low; personal thanks to Meg Beade, Jason Stowe, Marlowe Stowe. **MORA**: Carolyn Driedger, Daniel Williford, George Wuerthner; personal thanks to Ranger Mariah Radue. **CRLA**: Heather Wright; personal thanks to Elena, who was directing traffic on Rim Drive—we did it! **NOCA**: Katy Hooper, Jon Riedel. **YOSE**: Martin Hutten, Breezy Jackson, Sean Schoville, Greg Stock. **JOTR**: Rick Hazlett, Breezy Jackson, Chris McDonald; personal thanks to Ranger Marc Mahan. **DEVA**: Ali Ainsworth, Tyler Nordgren; personal thanks to Ranger Sarah Carter. **SEKI**: Danny Boiano, Sarah

Kramer, Nate Stephenson, Russell Winter. **LAVO**: Jessica Ball, Steve Buckley, Gregory Purifoy, Taza Schaming. **REDW**: Melissa Lockwood. **CHIS**: Katie Davis Koehn. **PINN**: Paul Johnson. **GRCA**: Jeanne Calhoun, Gregory Holm, Dave Thayer; personal thanks to Muriel Liebmann, Kelly Teale. **ROMO**: Kathy Brazelton, Miranda Redmond; personal thanks to Matt at Outback Saloon for being there when we really needed it. **ZION**: Walter Fertig, Dennis Terry, Curt Walker, Ann Winters. **YELL**: John Berini, Gill Geesey, Dan Tinker. **GRTE**: Leila Shultz, Dennis Terry; personal thanks to Kelsey Doherty, Kris Vrolijk. **GLAC**: Mark Biel, Tabitha Graves, ReBecca Hunt-Foster. **BRCA**: Sarah Haas, Tyler Nordgren, Eric Vasquez. **ARCH**: Annie Novak, Paul Lathrop, Ranger Mequette Gallegos, Finley at Woody's Tavern. **CANY**: Lee Ferguson, Tyler Nordgren; personal thanks to Ranger Robbie Anderson. **CARE**: Ed Myers, Tyler Nordgren. **SAGU**: Steve Buckley. **PEFO**: Marisa Acosta. **MEVE**: Julia Ponder. **CAVE**: Rod Horrocks. **GRSA**: Fred Bunch. **BIBE**: Steve Buckley, Jacqueline Richard. **BLCA**: Nickolos J. Myers, Jacqueline Richard. **GUMO**: Elizabeth Jackson, Jacqueline Richard. **GRBA**: Ronald Lanner; personal thanks to Officer J. Drew for hummingbird assistance. **CUVA**: Sonia Bingham, Doug Marcum, Arrye Rosser; personal thanks to Marcus Collins, Ranger Lisa Meade, Paul Schweigert. **INDU**: Erin Argyilan, Christine Gerlach, Kim Swift, Gia Wagner. **HOSP**: Shelley Todd; personal thanks to Officer Flint Stock, Ranger Lissa Allen, Ranger Kevin. **BADL**: Dennis Terry, Rachel Benton; personal thanks to Kellen Shaver, Laren Nowell, Paul Ogden, Ranger Jey. **THRO**: Becky Barnes, Daniel Swanson. **WICA**: JP Martin, Greg Schroeder. **VOYA**: Bryce Olson; personal thanks to Fame and Ashley at Almost Lindy's, Louise Whitehead. **ISRO**: Sarah Johnson, Liz Valencia. **GRSM**: Will Blozan, Lynn Faust, Kristine Johnson, Dennis Terry; personal thanks to Terry Bevino, Ranger Misty. **ACAD**: Ruth and Duane Braun, Catherine Matassa; personal thanks to Ranger Mackette McCormack, Ranger Alex. **SHEN**: Rolf Gubler, Dennis Terry; personal thanks to Gabriel Mapel, Ranger Christy, Kendra Bunnell. **EVER**: Michelle Collier, Antonia Florio, Jimi Sadle, Alan Scott. **MACA**: Leslie North, Rick Toomey, Michelle Verant. **BISC**: Gary Breman. **CONG**: Jonathan Manchester. **HAVO**: Janet Babb, Jessica Ferracane, Sierra McDaniel. **HALE**: Janet Babb, Patti Welton. **VIIS**: Laurel Brannick. **DRTO**: Meaghan Johnson. **NPSA**: Pua Tuaua. **DENA**: Robert Newman, Alex Webster. **GLBA**: Sarah Kramer, Ingrid Nixon. **WRST**: Caroline Ketron. **KATM**: Judy Fierstein. **KOVA/LACA/GAAR**: Tyler Nordgren. **ADDITIONAL FACT CHECKING**: Zak Martellucci.

# index

## A

Acadia (ACAD), 245–50
*Acer macrophyllum*, 10, 11
*A. rubrum*, 258
*A. saccharum*, 239
acorn barnacle (*Balanus glandula*), 13
acorn woodpecker (*Melanerpes formicivorus*), 59–60
adaptive radiation, 290
*Adelges tsugae*, 240
*Adiantum capillus-veneris*, 105
ʻae fern (*Polypodium pellucidum*), 283
*Aesculus flava*, 239
*Agapostemon* spp., 82
agaves, 154–56
*Aix sponsa*, 198
alligators, 260–61
alluvial fans, 50–52
Alpha Codes, 3
altitude sickness, 94–95
ʻāmaʻu fern (*Sadleria cyatheoides*), 283
*Amblysiphonella*, 190
American alligator (*Alligator mississippiensis*), 260
American bullfrog (*Rana catesbeiana*), 199
American crocodile (*Crocodylus acutus*), 260
American dipper (*Cinclus mexicanus*), 15–16
*Ammophila breviligulata*, 204–5
amygdules, 236
Ancestral Puebloans, 169, 172
andesite, 325, 326
Andromeda Galaxy, 138, 196
Anishinaabe (Ojibwe), 231
*Antilocapra americana*, 121–22
Appalachian Mountains, 118, 208, 237, 239, 243, 251, 255
*Aquilegia chrysantha*, 105
archaea, 111, 112
archaeoastronomy, 172
Arches (ARCH), 141–46, 159, 160, 161
*Archilochus alexandri*, 139
arctic disjuncts, 234
*Argyroxiphium grayanum*, 291
*A. sandwicense* subsp. *macrocephalum*, 290, 291
*Ariolimax columbianus*, 23–24
astronomy, 136–38, 172, 196, 304, 330–32
Atlantic dogwhelk (*Nucella lapillus*), 249
aurora australis, 332
aurora borealis, 332

## B

Badlands (BADL), 211–16
Badwater Basin, 49, 52
*Balanus glandula*, 13
bald cypress (*Taxodium distichum*), 259, 278
bald eagle (*Haliaeetus leucocephalus*), 198
banana slug (*Ariolimax columbianus*), 23–24
Bar Harbor Formation, 247, 248
barnacles, 13–14
bats, 174, 175, 269–70, 302–3
beach grass (*Ammophila breviligulata*), 204–5
Bear Gulch Cave, 84
bears, 252–54
beaver (*Castor canadensis*), 198, 199
bees, 82–83
bentonite, 218–19
Big Bend (BIBE), 181–84
big dipper firefly (*Photinus pyralis*), 242
bigleaf maple (*Acer macrophyllum*), 10, 11
big sagebrush (*Artemisia tridentata*), 123–24
bioluminescence, 241
birds, 168, 198–99, 206. *See also individual species*
Biscayne (BISC), 271–74
bison (*Bison bison*), 115–16, 223
black bear (*Ursus americanus*), 252–54
Black Canyon of the Gunnison (BLCA), 185–88
black cherry (*Prunus serotina*), 239
black-chinned hummingbird (*Archilochus alexandri*), 139
black crowberry (*Empetrum nigrum*), 235
black spruce (*Picea mariana*), 312, 313

black-tailed prairie dog (*Cynomys ludovicianus*), 227–28
black twinberry (*Lonicera involucrata*), 131
blisterpod, 272–73
Blue Cut Fault, 44
blue ghost firefly (*Phausis reticulata*), 241–42
bobby socks, 110
Bogachiel River Valley, 12
brachiopods, 90, 91, 190
Bridalveil Fall, 39
bristlecone pine (*Pinus longaeva*), 194–95
broadleaf cattail (*Typha latifolia*), 230
broad-tailed hummingbird (*Selasphorus platycercus*), 139
bromeliads, 262–64
Brule Formation, 214
Bryce Canyon (BRCA), 85, 133–40
*Bursera simaruba*, 258
*Buteo jamaicensis*, 170
butterflies, 36–38
butterwort (*Pinguicula vulgaris*), 235

## C

Cadillac granite, 247
*Calamagrostis scopulorum*, 105
calcium carbonate, 92, 190, 258, 299
*Candelaria pacifica*, 39, 40
Canyonlands (CANY), 141, 142, 145, 146, 159–62
canyon tree frog (*Hyla arenicolor*), 107–8
Capitol Reef (CARE), 147–52
carbon sequestration, 72
cardinal airplant (*Tillandsia fasciculata*), 263
cardinal monkey flower (*Mimulus cardinalis*), 105
Carlsbad Caverns (CAVE), 173–76
Carolina silverbell (*Halesia carolina*), 239
*Cassiopea frondosa*, 274
*C. xamachana*, 274
*Castor canadensis*, 198, 199
Catawba Indian Nation, 275
*Catopsis berteroniana*, 263
cattails, 230–32
caves, 84, 173, 176, 223, 265–66
cave swallow (*Petrochelidon fulva*), 174–75
*Centrocercus urophasianus*, 124
century plants, 154–56
*Cervus canadensis*, 86, 87
*Chamerion angustifolium*, 320–22
Channel Islands (CHIS), 77–80
Chardron Formation, 214
*Chelydra serpentina*, 199, 276
chert, 92
Chihuahuan Desert, 173, 181, 183, 189
chlorastrolite, 236
Cholla Cactus Garden, 43
chromium, 165
chytrid fungus, 108
*Cinclus mexicanus*, 15–16
Clark's nutcracker (*Nucifraga columbiana*), 68–69
Cliff Palace, 172
climate change, 36, 72–73, 124, 132, 195, 214, 264, 292, 299, 307, 313, 318
clodius parnassian (*Parnassius clodius*), 37
coast redwood (*Sequoia sempervirens*), 71, 72–73
coconut palm (*Cocos nucifera*), 294–95
*Cocos nucifera*, 294–95
collared lizard (*Crotaphytus collaris*), 167
collared peccary (*Pecari tajacu*), 157–58
Colorado Plateau, 100, 169
Colorado River, 98–100, 185
Columbian sharp-tailed grouse (*Tympanuchus phasianellus columbianus*), 124
common juniper (*Juniperus communis*), 134–35
common periwinkle (*Littorina littorea*), 249
concretions, 149
Congaree (CONG), 275–78
coral reefs, 298–300
*Corvus corax*, 86, 87, 145–46
cottonwood (*Populus deltoides*), 205
Crater Lake (CRLA), 25–28
creosote bush (*Larrea tridentata*), 53–54
crinoids, 90, 91, 92
crocodiles, 260–61
cross bedding, 102–3
*Crotaphytus collaris*, 167
cryptobiotic crust, 142–44
Crystal Cave, 55
cuckoo bee (*Townsendiella ensifera*), 82, 83

Cuyahoga Valley (CUVA), 197–200
*Cyanidioschyzon*, 111
cyanobacteria, 126, 142, 143, 282
*Cylindropuntia bigelovii*, 42–43
*Cyphoderris strepitans*, 124
cypress domes, 259

**D**

dacite, 325
dams, 15–16
Death Valley (DEVA), 49–54
debris flows, 18–20
decomposition, 21–22
Denali (DENA), 309–14
desert fan palm (*Washingtonia filifera*), 46
desert glue, 142
desert varnish, 160–61, 164
Diné (Navajo), 177
dinosaurs, 167–68, 184, 191
*Dipodomys* spp., 47–48
*Distichlis spicata*, 221
Douglas-fir (*Pseudotsuga menziesii*), 30–31, 73
driving, 6
Dry Tortugas (DRTO), 297–300
*Dubautia menziesii*, 291
dunes, 177–80, 201–6, 327

**E**

earthquakes, 44, 120
eastern box turtle (*Terrapene carolina*), 277, 278
eastern river cooter (*Pseudemys concinna*), 276, 278
elk (*Cervus canadensis*), 86, 87
Ellsworth schist, 247
Elwha River, 15–16
*Empetrum nigrum*, 235
*Enhydra lutris*, 79
*Entada gigas*, 273
epiphytes, 10–12, 262–64
*Eupithecia* spp., 285–86
Everglades (EVER), 257–64
extremophiles, 110–12

**F**

fall colors, 192
false dawn, 331
faults, 44–46, 118–20
feldspar, 187
*Festuca arundinacea*, 105
fireflies, 241–42
fireweed (*Chamerion angustifolium*), 320–22
fjords, 315
flying foxes, 302–3
fossils, 90–92, 126–28, 190–92, 211–16, 218, 219
*Fouquieria splendens*, 182–83
*Fratercula cirrhata*, 316–18
*F. corniculata*, 316–18
frigatebirds, 300

**G**

gabbro, 247, 248
Gates of the Arctic (GAAR), 329–32
gear, basic, 3
geosmin, 226
giant airplant (*Tillandsia utriculata*), 263
giant kelp (*Macrocystis pyrifera*), 78–80
giant sequoia (*Sequoiadendron giganteum*), 55, 56–58
glacial erratics, 34–35
Glacier (GLAC), 125–32
Glacier Bay (GLBA), 305–8, 315
glaciers, 18–20, 26, 29, 34–35, 62–64, 234, 305–7
glochids, 42
gneiss, 186, 187
goethite, 165
golden columbine (*Aquilegia chrysantha*), 105
gooseneck barnacle (*Pollicipes polymerus*), 13
gopher snake (*Pituophis catenifer*), 150–52
Grand Canyon (GRCA), 85–92, 185
Grand Staircase, 81, 101, 133
Grand Teton (GRTE), 117–24
grasses, 221–22
gray whale (*Eschrichtius robustus*), 74–76
Great Basin (GRBA), 193–96
great blue heron (*Ardea herodias*), 198, 199

greater sage-grouse (*Centrocercus urophasianus*), 124
Great Rift, 330
Great Sand Dunes (GRSA), 177–80
Great Smoky Mountains (GRSM), 237–44
greenish blue (*Plebejus saepiolus*), 37
Greenstone Ridge Lava Flow, 236
green sweat bee (*Agapostemon* spp.), 82
greensword (*Argyroxiphium grayanum*), 291
Guadalupe Mountains (GUMO), 189–92
gumbo limbo (*Bursera simaruba*), 258
Gunnison River, 185

## H

Haleakalā (HALE), 287–92
Haleakalā silversword (*Argyroxiphium sandwicense* subsp. *macrocephalum*), 290, 291, 292
*Halesia carolina*, 239
*H. monticola*, 239
*Haliaeetus leucocephalus*, 198
hamburger beans (*Mucuna sloanei; M. urens*), 273
hammocks, 258
hanging gardens, 104–6
Hawai'i Volcanoes (HAVO), 66, 279–86
*Helianthus petiolaris*, 178–80
helmock wooly adelgid (*Adelges tsugae*), 240
hematite, 148–49, 164, 165
*Hesperostipa comata*, 221, 222
Hetch Hetchy, 40
Hexie Mountains, 44
hiking, 7
Hoh River Valley, 12
hoodoos, 133
horned puffin (*Fratercula corniculata*), 316–18
Hot Springs (HOSP), 207–10
huckleberry (*Vaccinium* spp.), 131–32
hummingbirds, 139–40
humpback whales, 308
*Hydrogenobaculum*, 111
hydrogen sulfide, 66–67, 111
*Hyla arenicolor*, 107–8

## I

inclusions, 248
Indiana Dunes (INDU), 201–6
inflorescences, 179
*Iris versicolor*, 231
iron oxide, 148
Isle Royale (ISRO), 233–36

## J

Jackson Hole, 117, 118, 122, 123, 124
*Jamesia americana var. zionis*, 105
javelina (*Pecari tajacu*), 157–58
jellyfish, 274
Jones reedgrass (*Calamagrostis scopulorum*), 105
Joshua Tree (JOTR), 41–48
Joshua tree (*Yucca brevifolia*), 41
junipers, 134–35

## K

Kaibab Formation, 90, 92
kangaroo rat (*Dipodomys* spp.), 47–48
karst landscapes, 266, 268
Katmai (KATM), 323–26
Kautz Creek, 20
Kawuneeche Valley, 98, 100
kelp, 78–80
Kenai Fjords (KEFJ), 64, 305, 315–18
Kīlauea, 279, 282
Kings Canyon (SEKI), 40, 61–64
Kobuk Valley (KOVA), 327, 330
Krejci Dump Site, 200
krummholz, 96–97

## L

lahars, 18–20
Lake Clark (LACL), 328, 330
Lake Superior, 233, 234, 236
land
  protecting, 7
  recognition, 5
*Larrea tridentata*, 53–54
Lassen Volcanic (LAVO), 65–70
lava, 280–84
lava lichen (*Stereocaulon vulcani*), 283

Lechuguilla, 176
lichens, 39–40, 283
licorice fern (*Polypodium glycyrrhiza*), 10, 11
lightning, 89, 224–26
*Liriodendron tulipifera*, 239
*Lithobates sylvaticus*, 310–11
little bluestem (*Schizachyrium scoparium*), 221
little brown bat (*Myotis lucifugus*), 269
*Littorina littorea*, 249
*Lobaria oregana*, 10, 11
lodgepole pine (*Pinus contorta*), 113–14
Logan Pass, 129–30
*Lonicera involucrata*, 131
Lower Elwha Klallam Tribe, 15
*Lupinus rivularis*, 16

## M

*Macrocystis pyrifera*, 78–80
magnetite, 202–3, 206
magnificent frigatebird (*Fregata magnificens*), 300
Mammoth Cave (MACA), 265–70
manganese dioxide, 165
*Manicaria saccifera*, 272
Mars, 50–52, 148–49, 160–61
Mauna Loa, 279, 282
*Melanerpes formicivorus*, 59–60
Mesa Verde (MEVE), 169–72
*Metallosphaera*, 111
*Metrosideros polymorpha*, 283
Milky Way, 136, 196, 304, 330
*Mimulus cardinalis*, 105
moraines, 62
mountain dubautia (*Dubautia menziesii*), 291
mountain goat (*Oreamnos americanus*), 129–30
mountain silverbell (*Halesia monticola*), 239
Mount Desert Island, 245, 246
Mount Katmai, 323, 324
Mount Lassen, 65
Mount Mazama, 25, 26
Mount Rainier (MORA), 17–24
Mount Tehama, 70
*Mucuna sloanei*, 273
*M. urens*, 273
muskrat (*Ondatra zibethicus*), 199
*Myotis lucifugus*, 269

## N

narrowleaf cattail (*Typha angustifolia*), 230
National Park of American Samoa (NPSA), 301–4
Native peoples, 5. *See also individual tribes*
Navajo Sandstone, 102–3, 104
needle-and-thread grass (*Hesperostipa comata*), 221, 222
night skies, 136–38, 196, 304, 330–32
Nisqually River, 20
Norris Geyser Basin, 110–11, 112
North Cascades (NOCA), 29–32
northern blue flag iris (*Iris versicolor*), 231
northern lights, 332
northern red oak (*Quercus rubra*), 239
Novarupta, 324
*Nucella lapillus*, 249
*Nucifraga columbiana*, 68–69
nunataks, 64
*Nuphar lutea*, 199
nurse logs, 21–22

## O

oases, 46
octotillo (*Fouquieria splendens*), 182–83
ʻōhelo shrub (*Vaccinium* spp.), 282, 283
ʻōhiʻa tree (*Metrosideros polymorpha*), 282, 283
old-growth forests, 30–32, 72–73, 238, 240
Olmsted Point, 35
Olympic (OLYM), 9–16
*Ondatra zibethicus*, 199
Oneida Indian Nation, 251
O'odham Nation, 156
orange sulphur (*Colias eurytheme*), 37
*Oreamnos americanus*, 129–30
Oregon lungwort (*Lobaria oregana*), 10, 11
Oregon spikemoss (*Selaginella oregana*), 10, 11
otters, 79

## P


Pacific Ring of Fire, 29, 328
painted lady (*Vanessa cardui*), 37
Palmer's century plant (*Agave palmeri*), 154, 155, 156
palms, 294–95
Pangaea, 190, 243
Parry's century plant (*Agave parryi*), 154, 155, 156
pe'a fanua (*Pteropus tonganus*), 302
pe'a vao (*Pteropus samoensis*), 302
*Pecari tajacu*, 157–58
pegmatite, 186–87
periwinkle, common (*Littorina littorea*), 249
permafrost, 312–14
Permian mass extinction, 92, 190, 191
Petrified Forest (PEFO), 163–68
*Petrochelidon fulva*, 174–75
*Phausis reticulata*, 241–42
phenology, 132
*Photinus carolinus*, 241, 242
*P. pyralis*, 242
*Physa zionis*, 104
*Picea glauca*, 312
*P. mariana*, 312
Pinnacles (PINN), 81–84
*Pinus contorta*, 113–14
*P. edulis*, 188
*P. elliottii*, 258
*P. longaeva*, 194–95
*P. ponderosa*, 88–89
pinyon pine (*Pinus edulis*), 188
*Pituophis catenifer*, 150–52
*Plebejus saepiolus*, 37
poison ivy, 240
*Pollicipes polymerus*, 13
*Polypodium glycyrrhiza*, 10, 11
ponderosa pine (*Pinus ponderosa*), 88–89
*Populus deltoides*, 205
potholes, 162
powdery strap airplant (*Catopsis berteroniana*), 263
prairie dogs, 227–28
prairie sunflower (*Helianthus petiolaris*), 178–80
prickly saxifrage (*Saxifraga tricuspidata*), 234
pronghorn (*Antilocapra americana*), 121–22
*Prunus serotina*, 239
*Pseudemys concinna*, 276
*Pseudogymnoascus destructans*, 269
*Pseudotsuga menziesii*, 30–31
*Pteropus samoensis*, 302
*P. tonganus*, 302
Puerto Rico Trench, 296
puffins, 316–18
pug moth caterpillar (*Eupithecia* spp.), 285–86
pumice, 324–26
pyrite, 165


## Q


Queets River Valley, 12
*Quercus rubra*, 239
Quinault River Valley, 12


## R


*Rana catesbeiana*, 199
rattlesnakes, 150, 228
raven (*Corvus corax*), 86, 87, 145–46
red maple (*Acer rubrum*), 258
red rocks, 148–49
red-tailed hawk (*Buteo jamaicensis*), 170–71
Redwall Limestone, 92
red-winged blackbird (*Agelaius phoeniceus*), 199
Redwood (REDW), 71–76
reflexed wild pine (*Tillandsia balbisiana*), 263, 264
rhyolite, 113, 325
Rialto Beach, 14
Rio Grande, 181
riverbank lupine (*Lupinus rivularis*), 16
river otter (*Lontra canadensis*), 199–200
rock squirrel (*Otospermophilus variegatus*), 86–87
Rocky Mountain (ROMO), 93–100
Rocky Mountain juniper (*Juniperus scopulorum*), 134–35, 222
Rocky Mountains, 93, 215, 218, 224
Roosevelt, Theodore, 217
rufous hummingbird (*Selasphorus rufus*), 139–40

## S

*Sadleria cyatheoides*, 283
safety, 6–7
sagebrush, 123–24
sagebrush cricket (*Cyphoderris strepitans*), 124
Saguaro (SAGU), 153–58
saguaro cactus (*Carnegiea gigantea*), 150
salmon, 15
saltgrass (*Distichlis spicata*), 221
San Andreas Fault, 44, 45, 46
*Saxifraga tricuspidata*, 234
schist, 186, 187
*Schizachyrium scoparium*, 221
*Schoenoplectus tabernaemontani*, 231
Schott's century plant (*Agave schottii*), 154, 155, 156
sea beans, 272–73
sea coconuts (*Manicaria saccifera*), 272
sea hearts (*Entada gigas*), 273
sea otter (*Enhydra lutris*), 79
*Selaginella oregana*, 10, 11
*Selasphorus platycercus*, 139
*S. rufus*, 140
*Semicossyphus pulcher*, 80
Sequoia (SEKI), 40, 55–60
*Sequoiadendron giganteum*, 55, 56–58
*Sequoia sempervirens*, 71, 72–73
shatter zone, 246–48
sheephead fish (*Semicossyphus pulcher*), 80
Shenandoah (SHEN), 251–56
Shenandoah salamander (*Plethodon shenandoah*), 256
silicon dioxide, 165
silverbells, 239
silversword, 290–92
sinkholes, 266–68
slash pine (*Pinus elliottii*), 258
snails, 249–50
snakes, 150–52
snapping turtle (*Chelydra serpentina*), 199, 276, 278
softstem bulrush (*Schoenoplectus tabernaemontani*), 231
solution holes, 259
Southern Cross, 304
southern maidenhair fern (*Adiantum capillus-veneris*), 105
Spanish moss (*Tillandsia usneoides*), 263
spatterdock (*Nuphar lutea*), 199
spire tops, 57
sponges, 91, 92, 190–91
spyhopping, 74
squirrels, 87
*Stereocaulon vulcani*, 283
stromatolites, 126–28
sugar maple (*Acer saccharum*), 239
*Sulfobus*, 111
sulfur, 66–67
sunflowers, 178–80
Sun Temple, 172
synchronous firefly (*Photinus carolinus*), 241, 242

## T

*Tachycineta bicolor*, 198
Tahoe Glaciation, 63
Tahoma Creek, 20
tall fescue (*Festuca arundinacea*), 105
talus, 26, 255
  caves, 84
  slopes, 255–56
tank bromeliads, 262, 264
*Taxodium distichum*, 259, 278
teddy bear cholla (*Cylindropuntia bigelovii*), 42–43
Tenaya Lake, 35
*Terminalia catappa*, 273
*Terrapene carolina*, 277
*Testudinalia testudinalis*, 249
Teton Range, 117–20
Theodore Roosevelt (THRO), 217–22
thermal waters, 110–12, 207–10
*Thuja plicata*, 31–32
Thunderhead Sandstone, 243–44
thunderstorms, 89, 224–26
tide pools, 13–14, 249
*Tillandsia*, 262, 263, 264
Timbisha Shoshone, 49
Tioga glaciation, 34–35
tortoiseshell limpet (*Testudinalia testudinalis*), 249

*Townsendiella ensifera*, 82, 83
*Trachemys scripta*, 276
tree rings, 88
tree swallow (*Tachycineta bicolor*), 198
Triangulum Galaxy, 196
tropical almonds (*Terminalia catappa*), 273
*Tsuga heterophylla*, 31
tufted puffin (*Fratercula cirrhata*), 316–18
tulip tree (*Liriodendron tulipifera*), 239
Tuolumne Meadows, 36, 38
turtles, 199, 276–78, 297
*Tympanuchus phasianellus columbianus*, 124
*Typha angustifolia*, 230
*T. latifolia*, 230
tyre palm (*Coccothrinax alta*), 294–95

## U

upside-down jellyfish (*Cassiopea frondosa; C. xamachana*), 274
*Ursus americanus*, 252–54
Utah juniper (*Juniperus osteosperma*), 134–35

## V

*Vaccinium* spp., 131–32, 282, 283
*Vanessa cardui*, 37
Virgin Islands (VIIS), 293–96
Virgin River, 104, 106
volcanoes, 17–20, 25–28, 65–67, 70, 81, 109, 113–14, 218–19, 279–84, 287–89, 323–24, 328
Voyageurs (VOYA), 229–32

## W

*Washingtonia filifera*, 46
waterfalls, 243–44
waterpocket fold, 147
western hemlock (*Tsuga heterophylla*), 31
western red cedar (*Thuja plicata*), 31–32
wetlands, 230–32, 257
whales, 74–76, 308
whitebark pine tree (*Pinus albicaulis*), 69
white-nose syndrome (WNS), 269–70
white spruce (*Picea glauca*), 312, 314
white-throated sparrow (*Bombycilla cedrorum*), 198–99
wildfires, 220–22
wildlife
  habituation, 254
  protecting, 7
wild rice (*Zizania palustris*), 231
Wind Cave (WICA), 223–28
wood, petrified, 164–66, 219
wood duck (*Aix sponsa*), 198
wood frog (*Lithobates sylvaticus*), 310–11
woodpeckers, 59–60
Wrangell-St. Elias (WRST), 64, 319–22

## Y

yellow-bellied slider (*Trachemys scripta*), 276, 278
yellow buckeye (*Aesculus flava*), 239
Yellowstone (YELL), 66, 109–16
Yosemite (YOSE), 33–40
*Yucca brevifolia*, 41

## Z

Zion (ZION), 85, 101–8
Zion jamesia (*Jamesia americana* var. *zionis*), 105
Zion snail (*Physa zionis*), 104
zodiacal light, 331, 332
zooxanthellae, 298, 299
*Zygogonium*, 111